Sport Hydration:
A Synopsis on Concepts and Applications

Stephen John Shroyer M.D.

ISBN 978-0-557-14642-0

Table of Contents

Preface

The seed was planted for this book at the time I was director of the Athletic Nutrition Program for the Neptune School System, Neptune, N.J.. I have always been a local person for most of my life. I was born at Fitkin Hospital, Neptune, N.J.. I graduated Neptune Senior High School in 1971. I graduated Monmouth College in 1976 with a Bachelor of Science in Medical Technology. I attended medical school at Universidad Autonoma de Guadalajara, Jalisco, Mexico. The year of graduation was 1980. I did my post-graduate education at Jersey Shore Medical Center, Neptune, N.J.. I am Board certified in Pediatrics.

In the early 90s, I was hired to be the school physician for the Neptune School System. Part of my duties was to attend the home football games. This is where I met and became friends with the Head Coach, John Amabile. After 1 season of play, I discussed with the coach the need for a Nutrition Program to counter the fatigue during the latter half of the football game. John had been coaching Neptune for about 10 years prior to my entrance onto the scene. He was on the verge of winning a Championship for Neptune; the last being 30 years prior. So I went out and developed an Athletic Nutrition Program for the Varsity Football Team. I engineered the program with the following attributes: Carbo-loading meals: dinner and breakfast and pre-game hydration with vitamins and ginseng. The in house food company ran the program in the school. The Head coach made it mandatory to attend. With the program up and running, the feedback from the coaches and players were very positive. The meals and the hydration were well tolerated. Neptune went on to win 3 State Championships starting in 1995. The game was played in Giant Stadium, N.J. against the great Franklin High School. Neptune played flawless under the QB, Justin Cella. Coach Amabile became the greatest head coach of Neptune Football in the history of the school.

This book is a synopsis of the medical, engineering aspects, to the fluid portion of this program. It presents the full back-ground knowledge of the nutrients $Na+$, $K+$ and $H2O$. The specific details of my program are intertwined in this presentation. I present this material using many facets for assimilation and understanding of the material. You will see and use Diagrams, Tables, Outlines and pictures to portray the information in a short, concise manner. I purposefully kept the written information short without wordiness. I suggest you study the sections intently and go through the calculations completely in order to understand the material. You will note endnotes during the reading which look like this: [#]. Please refer to the end of the book for the reference. This will provide further reading on the subject. Figures, graphs, and diagrams will be noted in the text by a *# (alphabet letter). They will be found at the end of each chapter. Hopefully at the end, you will take with you an in depth knowledge of the material and if necessary, you can develop and mange your own fluid program.

Chapters 1-4 covers the science of the three nutrients. Chapters 5-7 explains the cell biology of the membrane and it's functionality. Chapters 8-11 describes the forces behind the cell's handling of the nutrients. Chapters 12 and 13 is the center of the book. This is where the engineering aspects are detailed. Chapter 14 : ADH and neurological control, Chapter 15: The cooling process, Chapter 16: Acclimatization and Chapter 17: Field Applications: Collegiate and
NFL.

This book is intended for para-medical and medical professionals interested in fluid   physiology with applications to sports medicine.

I would like to thank my family for their patience in this endeavor and to the great coach, John Amabile, for allowing me to be myself and for the opportunity gained in the field of sports medicine.

Chapter 1

Homeostasis: Control of the Internal Environment

Homeostasis *(1)* is the maintenance of a constant state, a steady state, in the internal environment. This is a basic fundamental process even at the level of the cell itself. From the cell to the whole organism, the compartments are divided into internal and external environments. At the level of the cell, there is intracellular and organelle compartments. For example, the cytosol (cytoplasm) and the mitochondria. This theme is consistent from the cell level to the entire organism. The human body has the Extracellular Fluid, E.C.F.; also called the milieu intérieur and the I.C.F., Intracellular Fluid.. **The I.C.F. is larger than the E.C.F**. The E.C.F. being smaller conserves matter and energy, easier to sense or monitor, and faster to bring about compensatory changes necessary to maintain homeostasis. This process is automatic and sequential in nature; making the human being an automaton *(2)*. In essence, the human body is on autopilot in maintaining the balance needed to sustain life. This is the involuntary operation of life that does not require a conscience effort. But a conscience effort is very closely linked to these processes. Sequential processing is the biochemical reaction steps, again providing conservation of matter and energy and the ability to provide feedback for proper control and maintenance of the target point. The target point is the normal range of operation. For example, the normal range in blood of $Na+$ is 135-145 mEq/L *(3)*.

In biology, an entity has traditionally been considered to be alive *(4)* if it exhibits all the following phenomena at least once during its existence:
- Growth
- Metabolism: consumption and excretion
- Motion
- Reproduction: procreation
- **Responsiveness**: sensation and reaction

The human body is made up of cells, the basic living unit, and the internal environment. The basic social network is as follows:

cell–>tissue–>organ–>organ systems–>organism

- **cell**=basic unit of life, eg. Myocardial cell.
- **tissue**=groups of cells of the same type, eg. Colony of Myocardial cells.

- **organ**=<u>groups of tissue</u> of the different type, eg. Heart=connective, myocardial, epithelial and nervous tissues.
- **organ system**=<u>groups of organs</u> of different type, eg. Cardiovascular system=heart, vessels and nerves.
- whole **organism**=human.

Each cell performs a special function; yet they all possess basic characteristics with one another. Most of the cells can reproduce, eg. Sperm cells can not.

The body fluid, the largest component of the human body, is divided into Intracellular, I.C.F. which is the largest compartment of fluid, and Extracellular, E.C.F.. The ECF is in constant motion and mixes thorough out the body by the force of **diffusion**, <u>kinetic motion</u> of molecules. Water molecules are in constant Brownian motion due to the presence of heat. Heat is energy. Energy can do work; in this case, create motion. Kinetic means motion. Motion is a form of kinetic energy created by the presence of heat. Diffusion is a passive process where molecules move from an area of high concentration to an area of low concentration, eg. The movement of water, electrolytes, glucose and etc. A differential gradient is created by the difference in concentration of solute across a <u>semi-permeable membrane</u>. This membrane creates the compartmentalization of living systems. The solvent is H_2O. The solute is Na^+ and K^+.

We will focus our attention on Na^+, K^+ and H_2O and the driving forces related to them.

There are differences between ICF and ECF:

Solutes

ICF	ECF	
K^+	Na^+	O_2
Mg^{+2}	Cl^-	Glucose
$O4-2$	HCO_3^-	Fatty acids
	CO_2	Amino acids

H_2O is found in both compartments. Na^+ and K^+ are mirror-image in functionality. The K^+ is much larger than Na^+ in atomic mass. K^+ depletion can be much more serious than Na^+ depletion, eg. K^+ depletion is usually associated with total body dehydration, I..C. F. + E.C.F.. This state is much more profound than other states of dehydration. Recovery is slower and prolonged, days instead of hours.

The major functional systems are acting together in the homeostatic mechanism

(5).

- The **circulatory** system, part of the E.C.F., exchanges constantly with the Interstitial compartment and eventually the Intracellular compartment.

- The origin of the nutrients in the E.C.F. are provided by the **respiratory**, **gastrointestinal**, **liver** and **musculoskeletal** systems.

- The removal of metabolic waste is managed by the respiratory and **renal** systems.

- The regulatory function is under the domain of the **nervous** and **endocrine** systems.

- Finally, the **reproductive** system generates a new being.

*1A Homeostasis of the Functional Systems

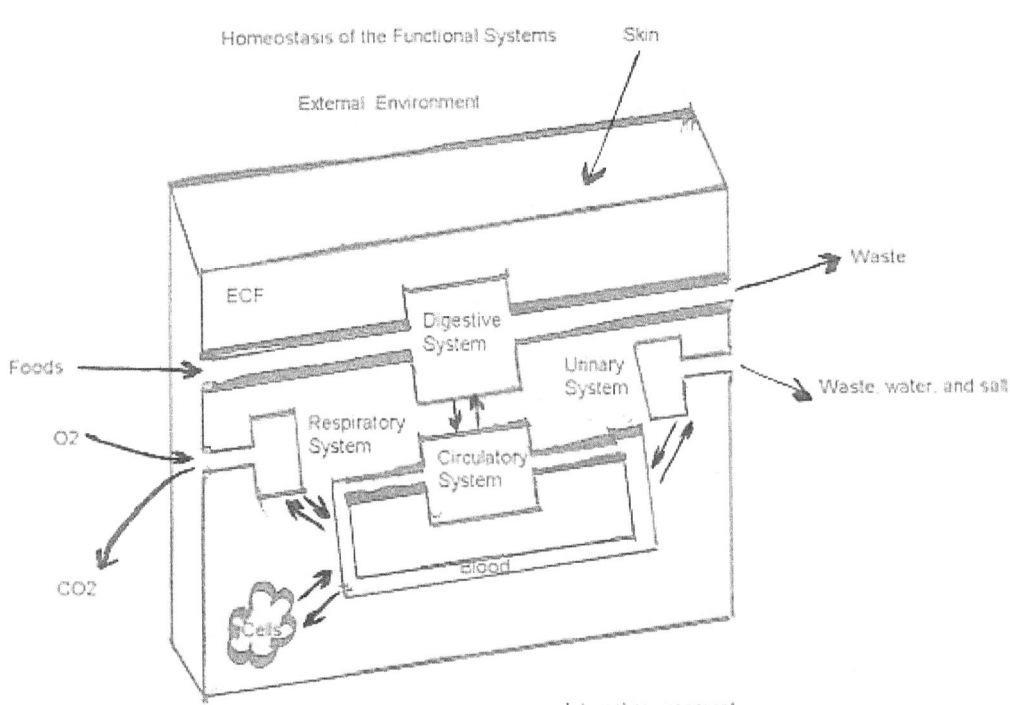

The human being can place himself in different environments and still maintain homeostasis, eg. football player in a football game.

The following is an explanation outline of the above diagram:

The external environment is the physical environment of the football game, the field, players, climate, water source, and etc..

The internal environment is the football player.

The skin separates the above 2 environments.

The internal environment is divided into Extracellular Fluid (ECF) and Intracellular Fluid.(ICF).

The fluids of the external environment, ECF, and ICF all differ in composition, eg. the Na+ content of a water bottle on a table at the game, Na+ in the blood of the football player, and Na+ content in a heart cell.

The main homeostatic systems are listed, Digestive, Respiratory, Urinary, and Circulatory. All interact with each other is this great interplay.

The cells are represented by themselves; they contain the ICF. Remember cells make up the various systems as described earlier.

Cells of organs exchange materials with each other via the internal environment, eg. Thyroid cell exchanging with a heart cell.

Blood is made of cells and plasma (the fluid component of blood).

The circulatory system is the main transporter of the internal environment. The heart is the pump and the vessels are the pipes

The Respiratory, Digestive, and Urinary systems exchange materials between the internal and external environments.

Let's take the case of Na+, this nutrient's path could be

      1. Ingestion:
                                a. CNS = The conscious sensation of thirst, so you
drink   b. Musculoskeletal =The act of swallowing
                                  c. Gastrointestinal=Digestion and absorption

2. Transportation:

     a. Cardiovascular system with dispersion by diffusion from ECF to ICF

     3. Concentration regulation: by

     a. **Osmosis**= water movement: universal (entire organism)

     b. **ADH** (Antidiuretic Hormone = water movement : Central Nervous System and Renal Negative Feedback mechanism

This homeostatic mechanism provides this nutrient for cardiovascular, nervous, muscular and thermoregulatory control and function.

So as you can see, that there are many types of control systems in the body. Systems can be in organs themselves, between organs and even in cells throughout the entire body; e.g. the genetic control system.

The control systems are 3 types in character:

     1. **Negative** feedback: Abnormal high is lowered. eg. High Na^+ - ADH - Low Na^+

     2. **Positive** feedback: Abnormal high goes higher. eg. Uterine contraction- Oxytocin- more Uterine contraction.

     3. **Adaptive** control (feed-forward): delayed negative feedback: Abnormal high is lowered in small increments. (refinement). eg. Strong Muscle movement - weaker muscle movement in increments.

The following is a table of Normal values in humans: *(6)*

E.C.F.

	Normal Value	Normal Range	Short -Term Nonlethal Limit Approximated
Na^+	142	138-146	115-175
K^+	4.2	3.8-5.0	1.5-9.0
H_2O*	60%*kg	45-75%**	(5-10-**15**%)-(<285 mOsm/L + **<135** mmol/L)

Unit:

Na^+ mmol/L

K+   mmol/L
H2O*   (%, mOsm/L, mmol/L)

* H2O is approximated as % of body weight (.6*kg). (0.6 x wt (kg))
Deficits of H2O is the same: ([.05,.1, .15]*kg).
Over hydration is evaluated and approximated by use of serum osmolarity (<285) and Na+ (<135).

** Depends on fat content (More fat less H20)
mMol/L=mEq/L.

Chapter 2

Water

The cell is divided into the nucleus and cytoplasm. A membrane separates each component. The substance make up of the cell is called protoplasm which contains 5 basic chemical entities: water, electrolytes, proteins, lipids, and carbohydrates.

H2O:

- principle fluid medium.
-
- largest concentrated chemical of the cell, except in fat cells.
-
- has a both dissolved and particulate entities associated with it.
-
- * place for **chemical reactions.**

H2O is an extremely important chemical compound. A review of it's chemistry is in order.

Matter is organized in the following social structure:

Subatomic particles (electron, proton, and neutron) –>**Atoms** (bonding) –> **Molecules** simple (single or small) or complex (multiple, large, complex, macro, polymer)–>**organelles**–>**cell.**

Chemical bonding (7) is important in the understanding of molecular behavior. It all starts with the atom. As stated, the atom consists of subatomic particles. The electron determines the atom's chemical reactivity.

- 1. H: Can occupy a total of 2 electrons in it's outer shell , It has only 1e in its outer shell; therefore it can accept 1 electron to fill it's outer shell.

- 2. O: Total 8, 6e outer; therefore it can accept 2 electrons.

Both have unfilled outer electron shells making them reactive.

*2A The Atomic Structure of the Elements of Water

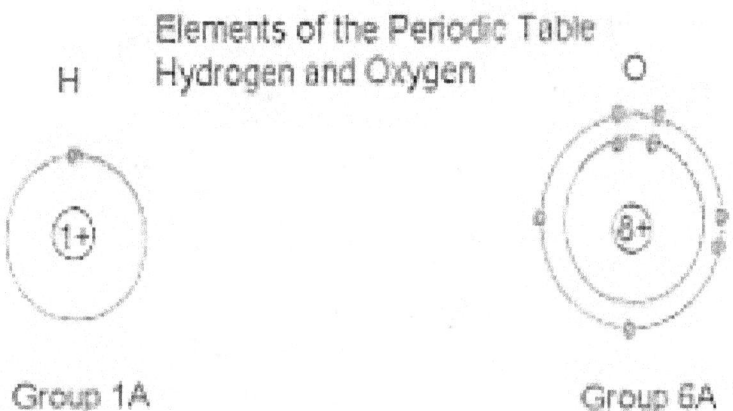

Explanation of the diagram:

The center circle with the #+ is the proton #. It is also the <u>nucleus</u> of the atom. There are equal # of neutrons as there are protons.

1+ =1 proton : nucleus = 1 proton + 1 neutron.

8+ = 8 protons: nucleus = 8 protons + 8 neutrons.

The outer rings are called electron shells.

The dots on the shell represent <u>electrons</u>.

The outer most shell of electrons are called **valence electrons which are reactive**. They can form chemical bonds. Atoms are more stable if their outer shell is filled with electrons. These atoms are called inert because they do not form chemical bonds.

Electrons prefer to be in pairs. Unpaired electrons are looking for a partner because they are **unstable**.

H = 1 shell that holds 2 electrons. Outer shell has 1 electron. Needs 1 electron to fill it's outer shell.

O = 2 shells that hold 8 electrons. Inner shell is filled. Outer shell has 6 electrons.
Needs 2 electrons to fill it's outer shell.
This form of diagram is called a **Lewis Dot structure** which depicts the outer valence
electron for better understanding of atomic chemical reactivity.

There are 5 ways in which atoms can react (*8)* with their characteristics:

1. Covalent bond: <u>Sharing</u> of e.
Strongest bond: 50-110 Kcal/mol.
Creates molecules: multiple atoms bound together.

2. Hydrogen bond: Sharing H atom: <u>Electrostatic attraction</u>:
Coulombs's Law
2nd strongest bond: 3-7 Kcal/mol.
Molecule to molecule attraction: Macro structure

3. Ionic bond: <u>Gain or lose e.</u>
2nd strongest bond: 3-7 Kcal/mol.
Creates molecules: salts: Crystalline Structure

4. van der Walls interaction: <u>Interaction of e clouds</u>.
Weakest bond: 1 Kcal/mol.
Molecule to molecule.

5. Hydrophobic interaction: <u>Interaction of nonpolar</u> substances.
3rd strongest bond:1-2 Kcal/mol.
Molecule to molecule: Saturated carbon
molecules.

Bond energy is the amount of energy to break a bond.

*2B The Chemical Bonds and their Interactions

Chemical Bonds and Interactions	bond energy kCal/Mol
Covalent bond : Sharing of electron pairs	50-110

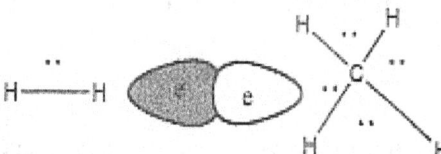

Hydrogen bond : Sharing of H atom	3-7

Ionic bond : Attraction of opposite charges	3-7

van der Waals interaction : interaction of electron clouds	1

Hydrophobic interaction : interaction of nonpolar substances	1-2

Explanation of the diagram:

The line joining the atoms is a covalent bond. **Covalent means sharing** a pair of electrons.
So the line is equal to 2 electrons, one electron from each atom.
C = 4 valence electrons.
O = 6
N = 5
Study the first chemical structure carefully.

In **sharing of the H** atom, **hydrogen bonding**, there is a partial charge on each atom which creates the electrostatic attraction between the two atoms.
H = partial + charge
O = partial - charge
This describes **polarity**, the unequal distribution of the electron (-) charge around the atom. The difference in the **electronegativity** of each atom is the determining factor.
N = partial - charge
The electron (-) of the O atom is attracted to the proton of the H (+).

The **ionic** attraction, **gain or lose of e,** is a much stronger polarity than H-bonding.
O = 7 valence electrons (1 more than 6). It has -1 **formal charge** (1 whole (-) charge).
N = 6 valence electrons (1 more than 5). It has a -1 formal charge (1 whole (-) charge).
H = 0 valence electrons (1 less than 1). It has a +1 formal charge (1 whole (+) charge).
H = 1 valence electron normally. Neutral or 0 charge.

van der Waals interaction is the interaction of the 2 or more **electron clouds**. (-) repels (-). This determines the spacial relationship among atoms in 3D space, molecule to molecule or within the same molecule. The cloud can vary in negativity creating a relative positivity which would create an attraction force: (+) attracts (-). Also (+) repels (+) for completeness sake.

Hydrophobic interaction is between **non-polar** substances, similar to van der Waals interaction.

Interactions are not formally chemical bonds. They are the interplay as described above between atoms.

Looking at H2O, we see that H2O is made of 2 atoms, H and O, held together by **covalent** bonds; water atoms share their electrons. Water molecule has covalent bonds, but the atoms share their electrons unequally, creating **polarity**, partial charges (*9*). O has a higher electronegativity than H and therefore has more electrons near it; giving it a partial (-) charge and H a partial (+) charge. This property gives water the ability to dissolve compounds; making water the chemical environment for the body's chemical reactions. If any compound can dissociate into an ionic specie, them it can associate with water.

*2C The Polar Covalent Bond of Water

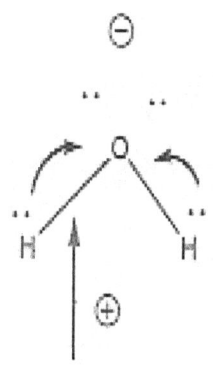

polar covalent bond

O is more electronegative than H, therefore it holds the electron pair more giving it a partial negative charge. The electron density is greater around the O atom. Electron density is the probability of finding the location of the electron in 3D space. Elements on the right side of the periodic table are more electronegative than the ones on the left side. The unpaired electrons are not part of the bond, but they are part of the molecule. They give the molecule it's geometry by occupying a position is space.

Explanation of diagram:

The positive and negative symbols represent the **dipole moment** of water or the polarity of the water molecule. In chemistry, this is usually represented by the Greek letter: Gamma ($\delta+/\delta-$). The head of the arrow is the negative area and the tail is the positive area. It depicts direction.

The line is the polar covalent bond which equals 2 electrons (electron pair).

The curve arrows signify that the pair is held closer to the O atom, because of it's higher electronegativity property.

The hydrogen bond (*10*) is an inter/intra molecular bond between molecules created by this polarity of water.

*2D The Hydrogen Bond of Water

H-bond between atoms of the same or different molecules

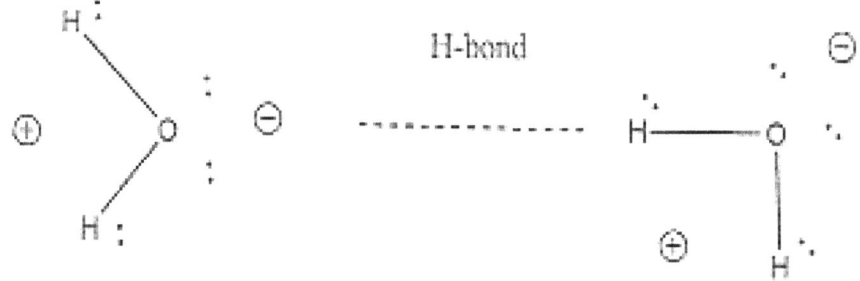

H-bond

Strongest in linear geometry as shown: O -- H

Explanation of the diagram:
These are 2 water molecules showing the H-bond.
Maximum bond strength is in linear geometry (180 degrees). It demonstrates the importance of direction and angle of approaching molecules for bonding or interactions to occur.

The H-bonding inside the same molecule could be between water or other polar function groups in the molecule, such as N – H ----- O = C.

Water: Structure + properties:
1. Tetrahedron shape (*11*)
2. Polarity
3. **H-bonding**

1. Solid (Ice) **less dense** than liquid (water): Air trapped inside the crystal. Ice floats on water. Ice insulates the water; keeps water below the ice from freezing.

2.High **Melting + Freezing** point temperatures: Temperature moderator:

3. High *****specific heat**: stabilizes temperature variations during heating and cooling.

4. High *****Heat of vaporization**: Cooling effect. (Liquid –> Gas)

5. High **Cohesive** strength: Water adheres to itself and other surfaces.

6. High **Surface tension**: The surface forms a strong mesh (difficult to puncture).

*2E The Geometry of Water

Tetrahedron geometry of H2O

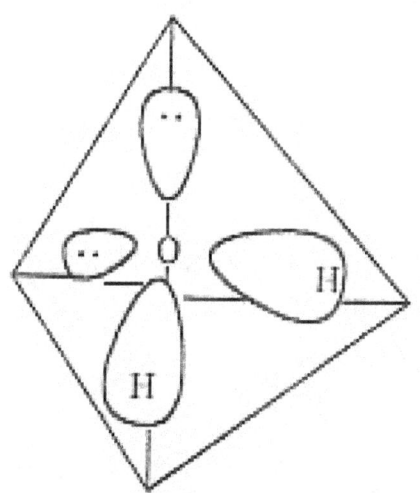

3D 4 sided triangle

Explanation of the diagram:

Imagine looking at a 3D object in a Molecular Orbital diagram.
The atoms are shown with the paired electrons (lone pairs).
The oval shapes are the sp^3 orbitals (hybrids). This is a probability surface demonstrating the area of high probability of finding the electrons. It is an approximation at best. The electrons are usually located near the nucleus. The optimum distance from the nucleus being dictated by Couloumb's Law.

Anomalous Properties of Water (Due to H-bonding) (*12*)

Specific Heat	4.18 J/(g*K)	Temperature Maintenance (prevent change)
Heat of Fusion	333 J/g	Thermostatic effect at freezing point
Heat of Vaporization	2250 J/g	Cooling effect (heat transfer)
Conduction of Heat		
Viscosity	10^{-3} N*s/m^2	Easy flow to equalize pressure
Dielectric constant	80 at 20 degrees C	Able to keep ions separated in solution
Surface Tension	$7.2*10^9$ N/m	Hard to penetrate surface

In summary, water acts as a thermal regulator and a glue-like substance. It fights the effects of extreme temperatures, a destructive biological result, and brings itself together, approximates substances, dissolves substances, associates with itself and other substances, shortens the distance for chemical reactions (increases the chances of chemical reactivity by creating a common chemical environment).

Chapter 3

Sodium

Ions (electrolytes) take part in **chemical reactions** and cellular **control** mechanisms. Sodium has the symbol **Na** $\left(\textit{Natrium},\text{ from Arabic }\textit{natrun},\right)$. It's atomic number is 11. Atomic mass of 23 amu (atomic mass units). Common oxidation number of +1. +1 means it gives up 1 electron (-1). Therefore it is an **ion**.

Shell filled # of e: (*13*)
Na: Total 11: 1 e outer
Na has <u>unfilled outer shell making it reactive</u>.

3A The Atomic Structure of Sodium

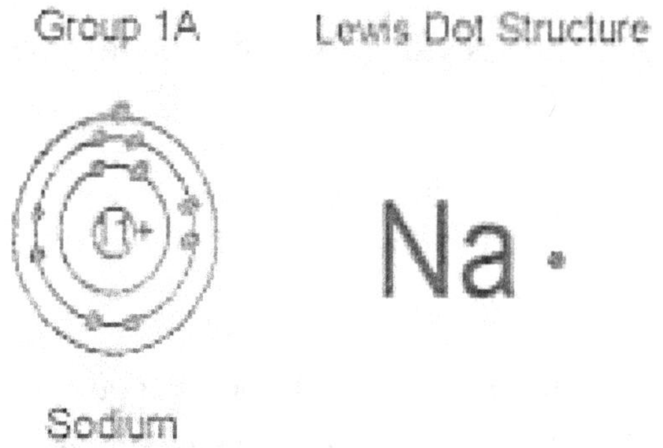

Group 1A   Lewis Dot Structure

Na ·

Sodium

Sodium is soft, silvery white, highly reactive element. It is a member of the alkali metals, group 1 (1A). (*14*)

*3B Sodium as a Solid

Sodium quickly oxidizes in air and is violently reactive with water. It must be stored in kerosene or mineral oil.

It is found in great quantities in the Earth's oceans as NaCl (common salt). Sodium is also found in many minerals. It is essential for animal life It is classified as a "dietary inorganic macro-mineral."

NaCL and KCL are made of 3 atoms, Na, K, CL, held together by ionic bonds; salt atoms gain or lose their electrons. CL gains 1 electron and Na loses 1 electron; each creating a filled outer shell. This stabilizes the bond (*15*)

*3C The Ionic Bond Formation in the Atomic Structure Format

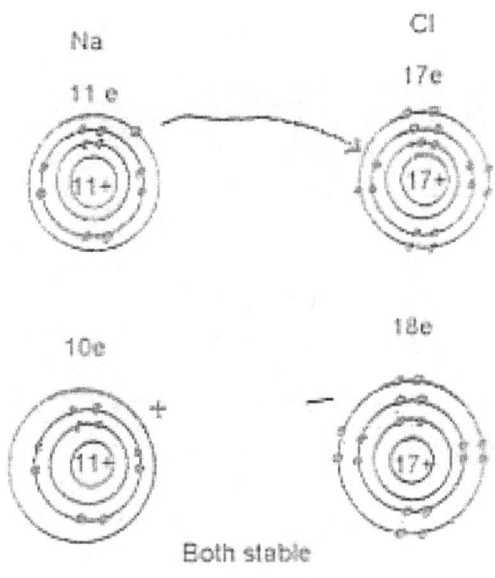

Na

11 e

Cl

17e

10e

18e

Both stable

Ionic bond fomation

Properties of Na+ (*16*)

Specific Heat   1.22 J/(g*K)
Heat of Fusion   113.098J/g
Heat of Vaporization   4237 J/g
Thermal Conductivity   142 W/(m*K) at 300 K
Physiologic importance: Water metabolism
                                  Endocrine System
                                  Nervous System
                                  Cardiovascular System

Summary: Na+ is very important in **Osmosis**, the force behind the movement of water, and in the **electrical-mechanical** and **electrical-communicative** systems of the body, example: muscle, nerve, and hormone glands.

Chapter 4

Potassium

Potassium is an **electrolyte (ion)**. It has the symbol **K**. It comes from the word kalium (Latin: kalium). It's atomic number is 19 and has an atomic mass of 39. It was first isolated from potash ((plant ashes). It is a member of Alkali metal: Group 1 (1A). Common oxidation number is +1.

Shelled filled # of e: K: Total 19: 1 e outer (*17*)
K+ has a unfilled outer shell making it reactive.

*4A The Atomic Structure of Potassium

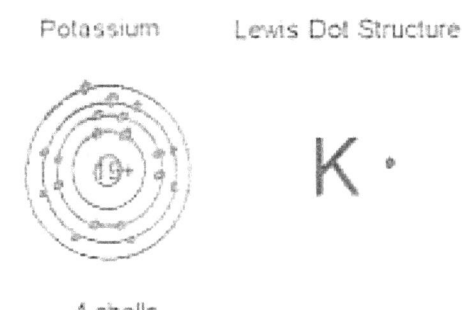

Potassium   Lewis Dot Structure

K ·

4 shells
1 valence electron
larger than Na , therefore different
same chemical reactivity as Na

Potassium is a soft silvery-white metallic alkali metal that oxidizes rapidly in air and is very reactive with water. The heat of the reaction could ignite the H release and create a flame (*18*).

*4B Potassium as a Solid

In nature, potassium occurs only as ionic salt. It is dissolved in seawater, and part of many minerals. It is present in plant and animal cells. Fruits have a high content of K.

Na and K are similar chemically. And therefore their biological importance is similar. They have been partitioned to accomplish their significance.

It is considered a "dietary inorganic macro-mineral."

Properties of K+ (*19*)

Specific Heat	.757 J/(g*K)
Heat of Fusion	59.59 J/g
Heat of Vaporization	2042.7 J/g
Thermal Conductivity	102.5 W/(m*K) at 300 K
Physiologic Importance:	Water metabolism
	Endocrine System
	Nervous System
	Cardiovascular System

Summary: Note the similarity to Na+. K+ is intracellular and Na+ is extracellular. Their functions are the same. They are the dynamic duo. They dance very well together.

Chapter 5

Membrane Lipids

Lipids are macromolecules which are insoluble in water. The biological functions of the lipids are diverse. The one in which we will study is the structural lipids of the cell membranes. The lipids create a bi-layer, which acts as a barrier to the passage of polar molecules and ions. The membrane compartmentalizes the cell itself; as well as, the cell as a whole from the Extracellular Fluid . The membrane is a very thin film of lipid and protein, about 5 nm thick (20). They are held together by non-covalent interactions. It is in a fluid state and dynamic in nature.

The RBC membrane is a good example of a membrane to study. With the electron microscope, we can see a light band and a dark band on cross section (side view). The light area is sandwiched between to dark (outer) areas. This is the bi-layer structure of the membrane (21).

*5A The Light and Dark Bands of the Bi-layer

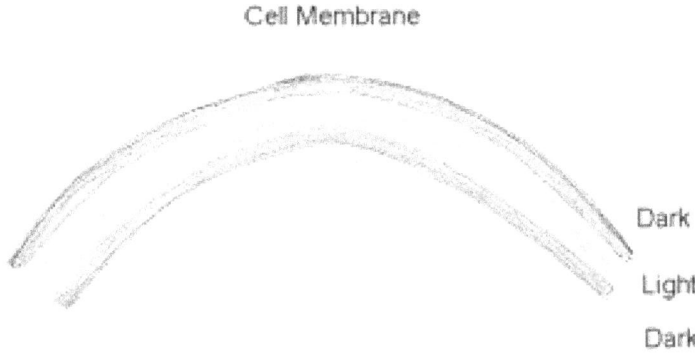

Cell Membrane

Dark

Light

Dark

Each half of the bi-layer is 1 molecule of the lipid. The layer is 2 molecules thick. The Dark band represents the polarized head of the lipid. The Light band represents the fatty acid chain of the lipid. They have different colors under the electron microscope due to their chemical makeup (22). Also seen in the bi-layer is the crossing of protein molecules.

22

*5B The Lipid Symbol

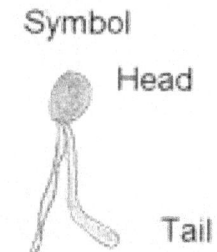

Symbol

Head

Tail

*5C The Lipid Bi-layer

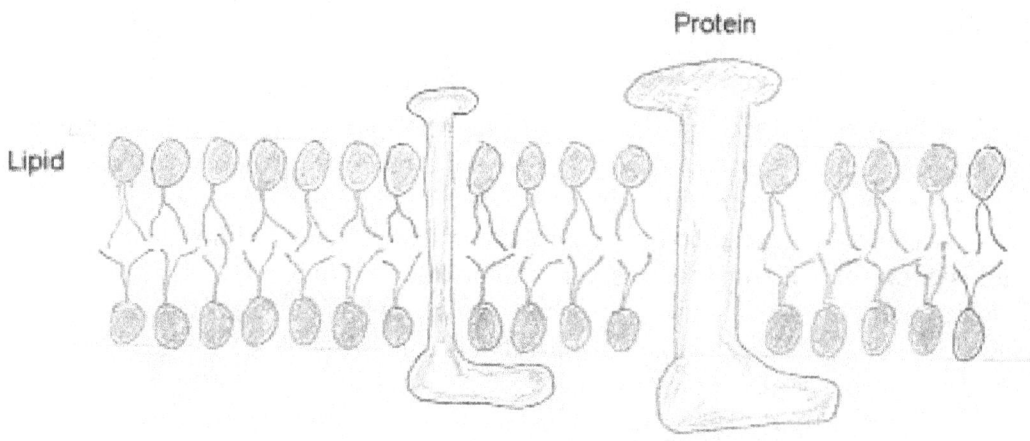

In a 3D drawing *(23)*, with all 3 axes (x,y and z), you get a real sense as to the true structure of the lipid bi-layer. The thickness as mention is only 5 nm (5 nm=$5*10^{-9}$) meters).

*5D 3D Bi-layer

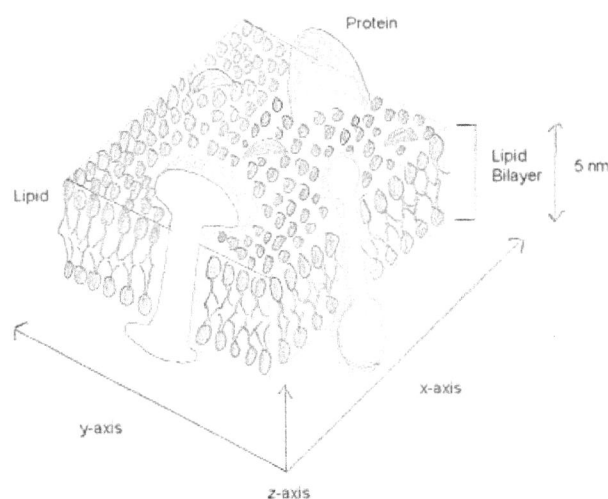

The most abundant of the lipids is the phospholipids (*24*). They are amphipathic or amphiphilic. This means that they possess a polar head group and a hydrocarbon tail. The polar head group is hydrophilic, it likes water. The tail is hydrophobic; it does not like water. The head is made of 3 molecules combined: Choline + Phosphate + Glycerol. The tail is 2 long chains of fatty acids, one with a kink in it (a bend). You think in terms of dissolution of the chemical: head dissolves in water while the tail dissolves in oil. A good example of a phospholipid is Phosphatidylcholine. At a pH of 7.4 (slightly alkaline), the head is ionized (charged/polarized).

*5E Phospholipid Schematic

Schematic

Phosphatidylcholine   (Phospholipid)

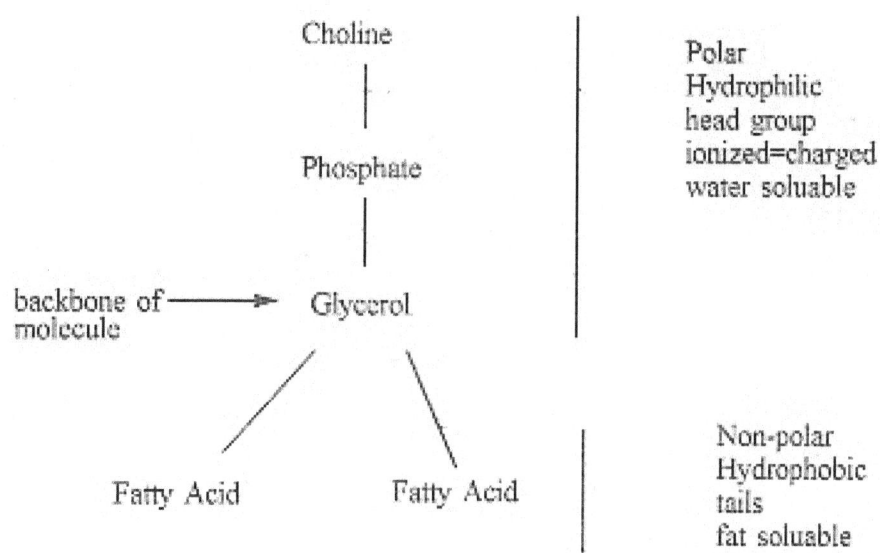

Choline — Phosphate — Glycerol → backbone of molecule; Fatty Acid, Fatty Acid

Polar
Hydrophilic
head group
ionized=charged
water soluable

backbone of molecule → Glycerol

Non-polar
Hydrophobic
tails
fat soluable

At a pH of 7.4 (normal body pH), the 2 ends of the molecule
will separate into their preferred soluable states.
pH is the log of the [H+]. In otherwords, the acid level of the body.
At 7.4, the body is slightly basic.

Formula (*25*)

Choline: CH2-CH2-N+-(CH3)3   (note the (+) charge)
Phosphate: PO4(-)   (note the (-) charge)

Cis: The 2 Carbon atoms associated with the double bond have a Hydrogen
oriented on the same side.

```
            H  H
            -C=C-
                  H
Trans:      -C=C-
                  H
```

Kink=determines packing, stacking of molecules. Prevents freezing (crystallizing)

*5F Phospholipid Formula

Formula

```
                        2HC ———— N+(CH3)3
                         |
                        CH2
                         |
                         O
                         |
                   O == P ———— O-
                         |
                         O
                         |
        2HC ———— C ———— CH2
         |       H       |
         O       O
         |       |
         C=O     C==O
         |       |
        CH2     CH2
         |       |
        CH2     CH2
       2HC     2HC
         |       |
        CH2     CH2
         |       |
        CH2     CH2
         |       |
        CH2     CH2
         |       |
        CH2     CH2
         |       |
        CH2     CH         ← Cis-double bond
         |       ‖            causes kink in chain
        CH2     CH            prevents freezing
         |       |
        CH2     CH2
       2HC     2HC
         |       |
        CH2     CH2
         |       |
        CH2     CH2
         |       |
        CH2     CH2
         |       |
        CH2     CH2
         |       |
        CH2     CH2
       3HC     3HC
```

van der Waals radii space-filling model (*26*)

Spheres=electron clouds
3-D

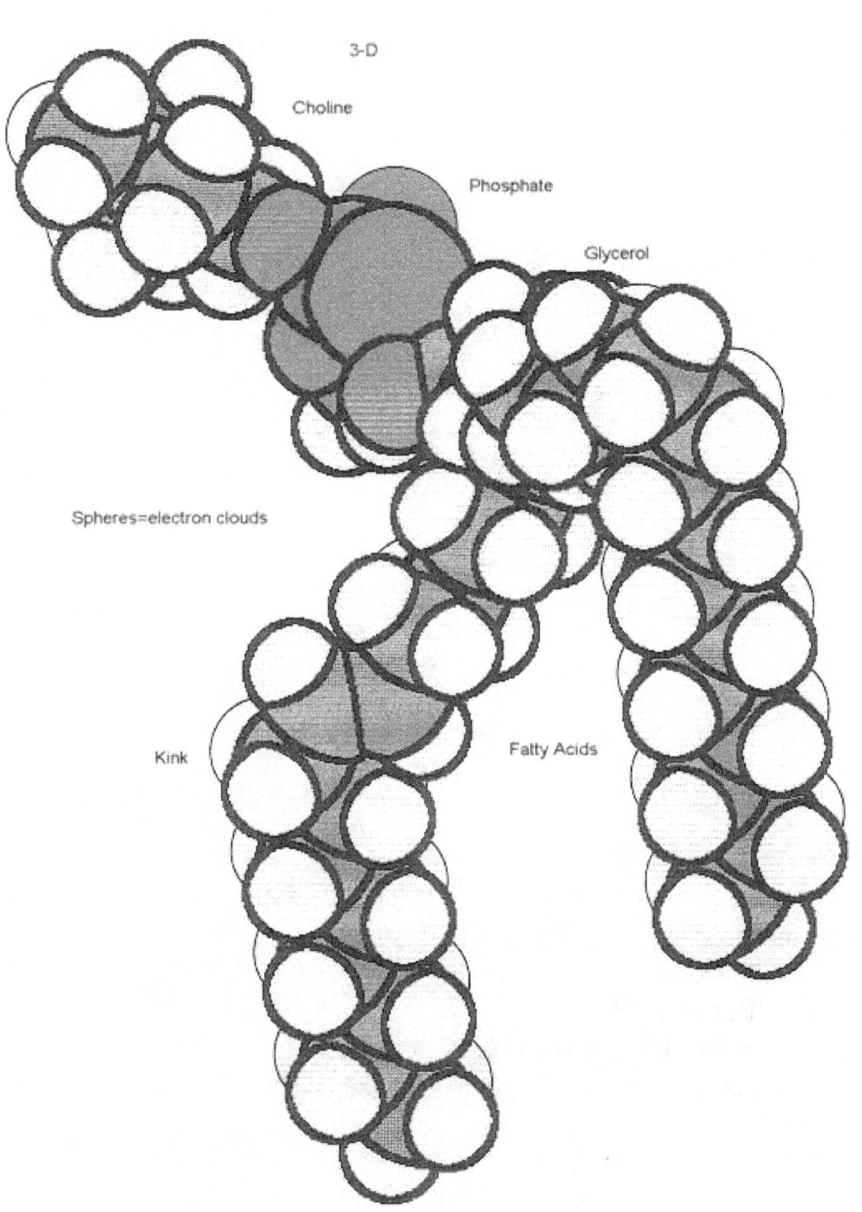

*5G Phospholipid van der Waals Structure

It is the shape that determines the structure of the lipid aggregate. Phospholipids are cylindrical and form bilayers, not micelles. They also seal themselves when torn.

Bi-layer=sheets
Micelles=spheres

Besides shape, it is thermodynamically directed, most stable with least energy. Other forces that determine the formation are: Hydrophobic interaction
van der Waals interaction
H-bonding

*5H 3D Lipid Motion

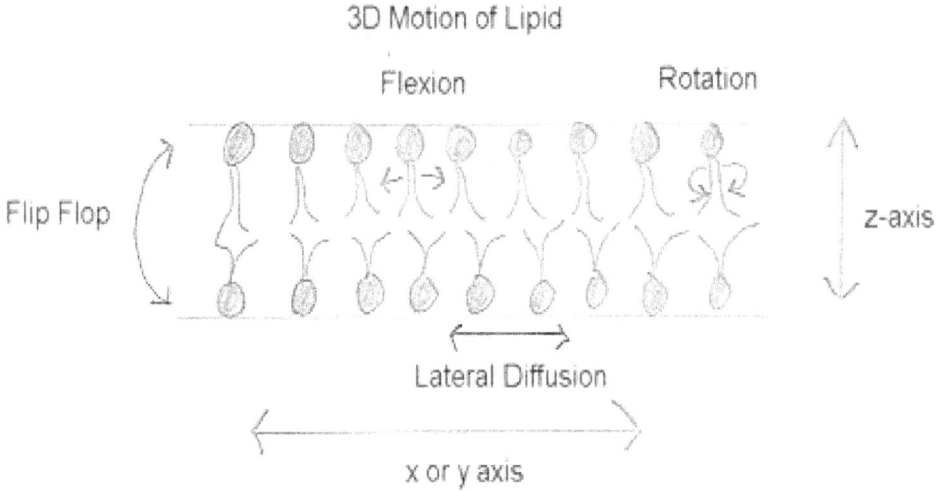

The bi-layer is considered a 2D fluid. There are 4 types of motion a lipid molecule can make: 1. Lateral Diffusion (most common) (*27*)
2. Flexion
3. Rotation
4. Flip-Flop (rare)
The first 3 is movement along the x + y axes. The 4th is along the z axis.

Fluidity: Phase transition: Liquid –>gel (freezing point)
Lowers the phase transition temperature:
Shorter Hydrocarbon chain (lowers hydrophobic

interaction)   Increase kinks=double bonds (decrease packing)
                    Cholesterol (increase rigidity of first 3 methylene groups
of hydrocarbon chain) (*28*)

*5I Steroid Ring Stiffened Region

1/2 bilayer (3nm)

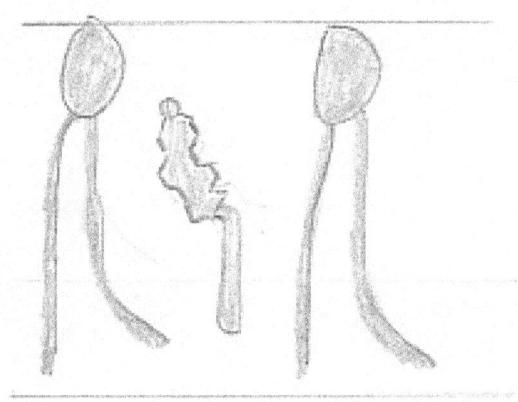

Steroid Ring Stiffened Region

Methylene group: -CH2-   (*29*)

Polar Head: Hydroxyl Group: -OH

Steroid: planar and rigid

Hydrocarbon tail

*5J Cholesterol Symbol

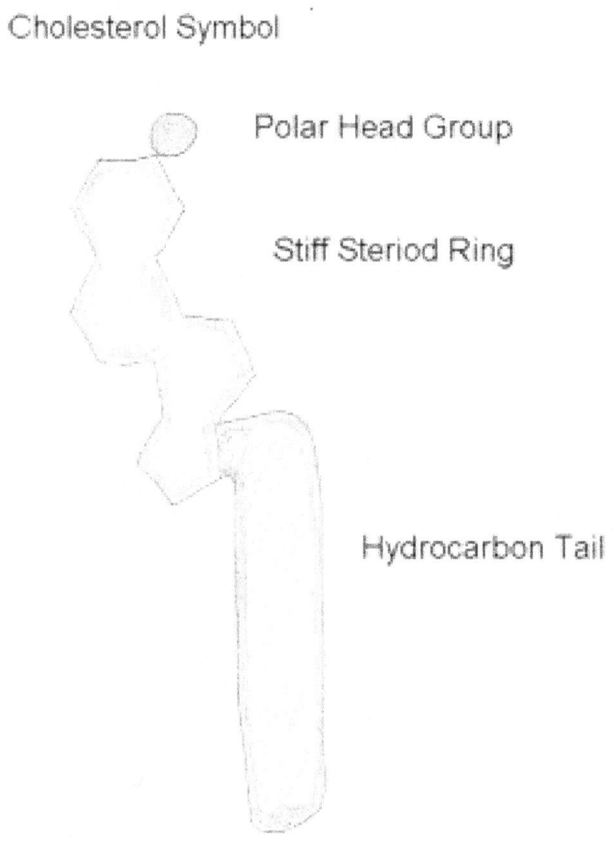

Cholesterol Symbol

Polar Head Group

Stiff Steriod Ring

Hydrocarbon Tail

*5K Cholesterol Formula

Formula

Cholesterol

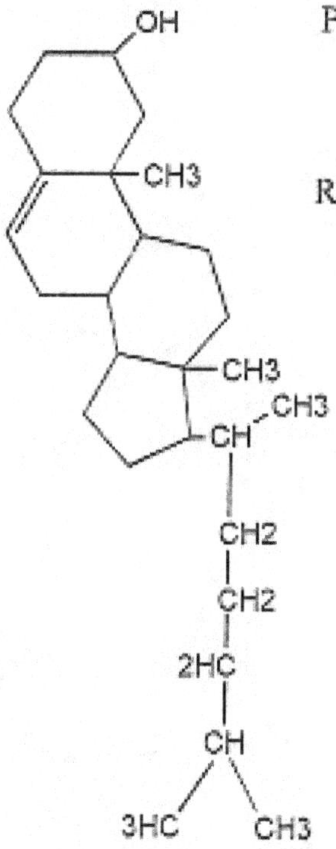

OH   Polar Head Group

CH3   Rigid planar steroid ring

CH3
CH3
CH

CH2
CH2   non-polar hydrocarbon tail
2HC

CH
3HC   CH3

*5L Cholesterol Space-Filling Model

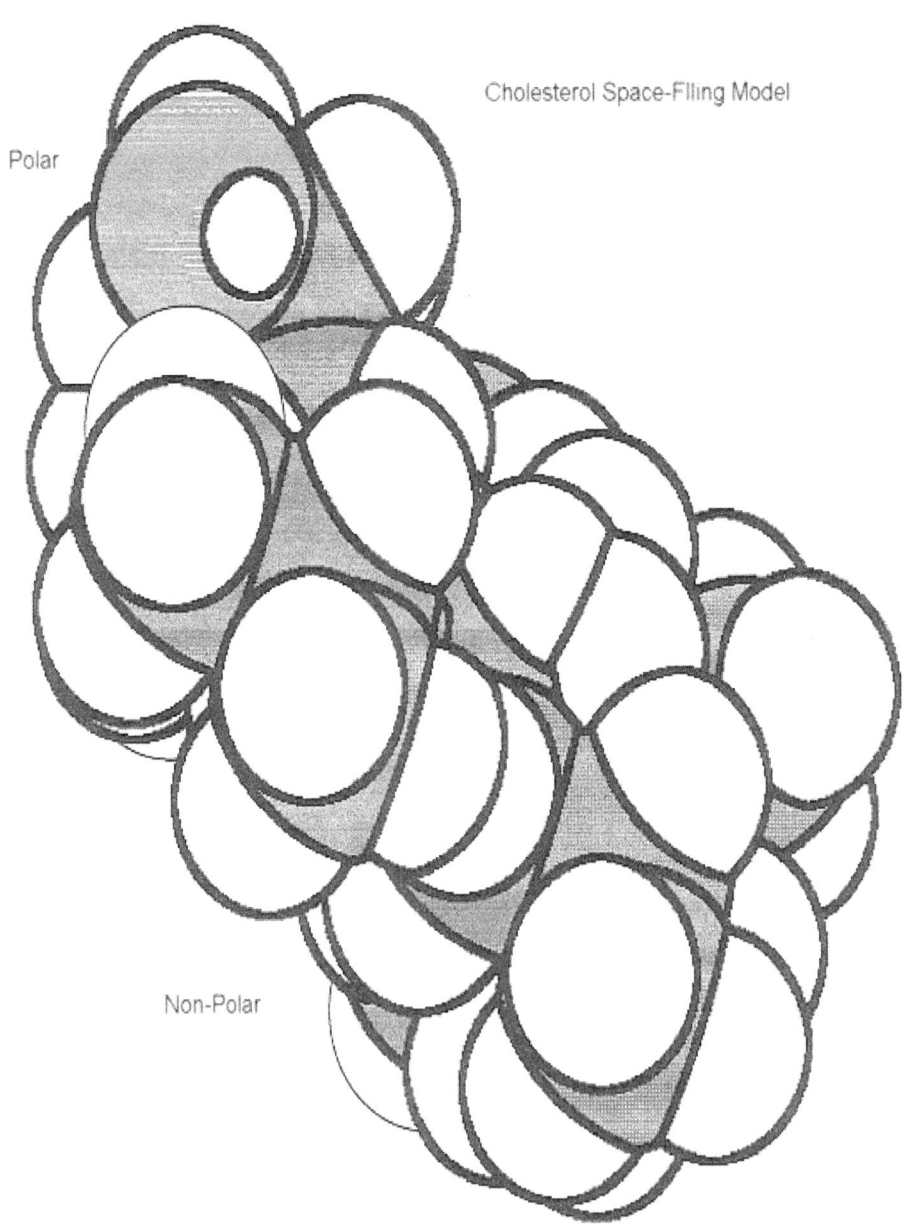

Cholesterol Space-Filling Model

Polar

Non-Polar

Asymmetry: Presence of different types of membrane lipids in both leaflets. Adds to versatility of functionality (*30*).

Phosphatidylcholine (% of Total Lipid by Weight)

Red Blood Cell Plasma Membrane	17
Liver cell	24
Myelin	10
Mitochondrion	39
Endoplasmic Reticulum	40
E. Coli bacterium	0

Summary: All cell and cell organelles possess membranes.
Lipids are part of membranes which act as a barrier to water and water-soluble solutes.
Lipids provide structure and function to cells.

Chapter 6

Membrane Proteins

The membrane function is provided by the proteins. As mention, the lipid bi-layer provided the basic structure of the membrane. The amount of protein depends on the cell or organelle function. There are 8 ways in which the protein attaches to the membrane (*31*).

The 2 main associations are the integral and peripheral proteins. Integral means that the protein is attached to the membrane through covalent bonding, H-bonding, hydrophobic interactions or van der Waals interactions. Peripheral means that the protein is not attached directly with the membrane, but is attached to other proteins which are attached to the membrane. They are attached to the anchor protein by H-bonding, hydrophobic interactions and van der Waals interactions.

Of the integral protein type, transmembrane protein pass completely through the membrane. They have one end of their molecule in the cytosol and the other end in the E.C.F.. The most important one that does this is the Alpha-helix type protein. This protein is a long chain protein with a counter-clockwise rotation about it's long axis. Kind of like a right-handed corkscrew. You take your right hand and put your thumb pointing up. Make a loose fist; the rest of the fingers point in the direction of the rotation of the protein. Now slowing raise your hand, you are tracing the backbone, the peptide bonds of the amino acids, of this long chain protein. The Alpha-helix can be a single pass or a multiple pass. The single Alpha-helix protein passes through the membrane once. The attachment is by a fatty acid. The protein is covalently bonded to the glycerol part of the fatty acid. The fatty acid chain attaches to the membrane by hydrophobic interaction.

Protein – Glycerol – Fatty Acid – Fatty Acid of the Membrane

In the multiple pass the chain bends on each side of the membrane to pass through the membrane again. It maintains it's orientation as it passes through the membrane multiple times. Meaning, the right-handed corkscrew orientation is maintained.

*6A Single and Multi-pass Proteins

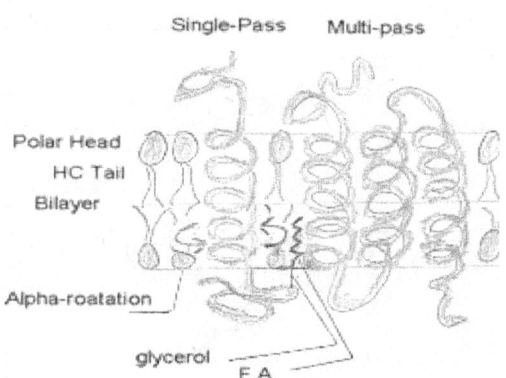

There other ways integral proteins attach to the membrane. Some attach with a fatty acid, this is done by covalent bond. An atom in the Glycerol part of the fatty acid is attached by covalent bond to the protein. Then the fatty acid attaches to the membrane by hydrophobic interactions. Remember the bilayer is essentially fatty acids in the interior of the membrane (the light band). Oil dissolves oil.

Protein – Glycerol – Fatty Acid – Fatty Acid of the Membrane

*6B Fatty Acid Attachment

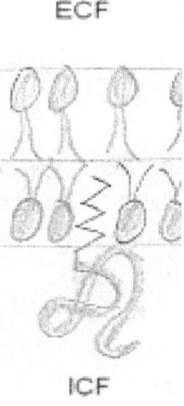

An Oligosaccharide, a small chain carbohydrate, could attach the protein to the membrane by a covalent bond.   The carboxyl end of the protein is phosphorylated. The oligosaccharide is attach to this phosphate. At the other end of the saccharide, this is phosphorylated. Here the Glycerol molecule of the fatty acid is attached and then the fatty acid chains connect the whole complex to the membrane. The fatty acid attaches to the membrane by hydrophobic interactions.

Protein-COOH – PO4- -- Oligosaccharide – PO4- – Glycerol – Fatty Acid – Fatty
        Acid of the Membrane

The Phosphate + Glycerol + F.A. is called a phospholipid.

*6C Oligosaccharide Association

Can you identify the phospholipid?

Other proteins are of a different mosaic type. Instead of forming an Alpha-helix, a secondary structure, they form a beta-sheet made from beta strands of protein chains. These take the shape of a barrel in 3D. The barrel passes through the membrane in a single pass. The barrel interacts directly with the membrane; attaches through H-bonding, hydrophobic interaction and van der Waals interaction.

Protein barrel – Polar and non-polar sections (fatty acid) of the membrane.

*6D Beta-Barrel Motif

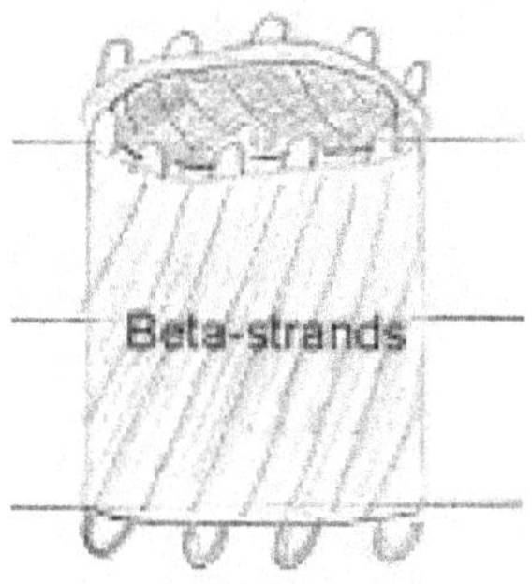

Beta-barrel

Beta-strands

The next association is the monolayer Alpha-helix. The Alpha-helix protein inserts itself into 1 side or ½ of the bilayer directly. The attachment is obtained through the H-bonding, hydrophobic interaction and van der Waals interaction. The right-handed corkscrew orientation is maintained. The amino acids of the protein have side groups which are polar and non-polar. These groups enable this process to occur.

Alpha-helix protein (polar and non-polar side groups of amino acids) – Polar and non-polar sections (fatty acid) of the membrane.

*6E Monolayer Motif

monolayer

The last association if the peripheral type. Proteins associate with other proteins by H-bonding, hydrophobic interaction and van der Waals interaction. 1 protein is attached to the membrane by one of the above mechanisms. The other protein is associated with it by what I just mentioned. The peptide bonds with their ability to H-bond and the side chains, polarized and non-polarized, drives these associations.

Protein – Protein (attached to membrane)

*6F Protein to Protein Association

Summary Outline: 1. Integral (Attached to the membrane)
                 Transmembrane (a protein that crosses the lipid bi-layer) (*33*)
                         1.Single-pass   a (alpha) helix
                                         a helix: right-handed corkscrew (counter-
clockwise rotation)   2. Multipass   a (alpha) helix
                         3. Fatty acid (attached by a covalent bond)
                         4. Oligosaccharide (small chain carbohydrate and attached
by a covalent bond)   5. Phospholipid (attached by a covalent bond)
                         rolled-up (*32*)
                         6. Beta Sheet
                         7. Monolayer (Alpha-helix in 1 layer of bi-layer)
                 2. Peripheral (Not attached to the membrane)
                         8. Protein to Protein (non-covalent interactions)

The peptide bond is the backbone to all proteins. It is planar (flat) in nature.

```
        |
    C=O
        |      Peptide bond
    NH
        |
```

View: 3 a.a. polypeptide on it's side being viewed down it's longitudinal axis of backbone.

*6G Planar Protein Backbone

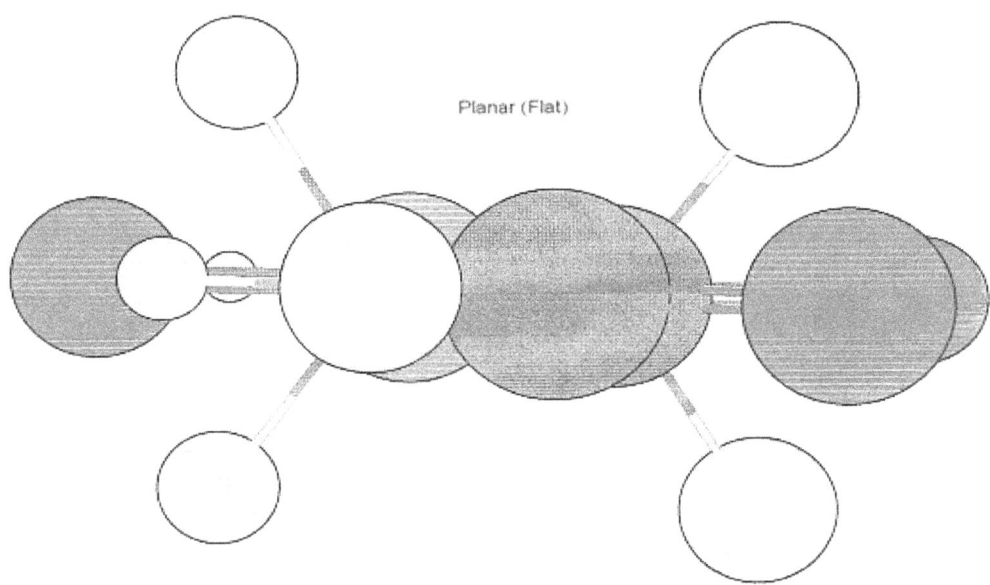

Planar (Flat)

peptide bonds are polar (H-bond with each other)

Piptide bond (H-bonding)

-C=O- - - - H-N-   (No water in bilayer)

non polar side chain (non-ionized /not charged)

*6H H-Bonding between Amino Acids of a Peptide Chain

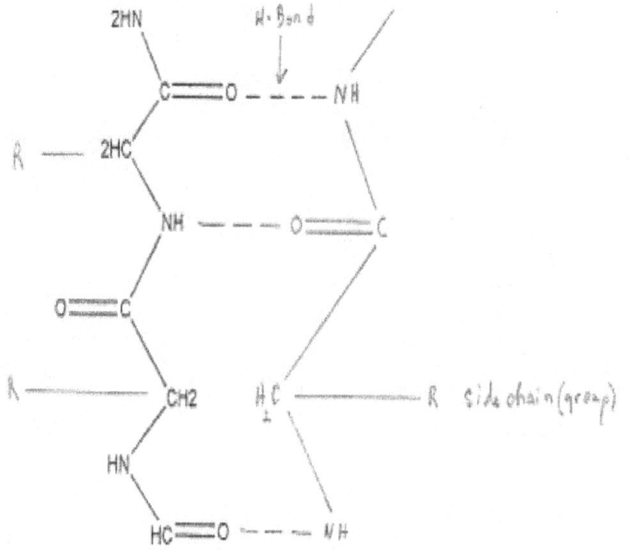

Let's examine a 4 amino acid polypeptide (34):

*6I 4 Amino Acid Peptide

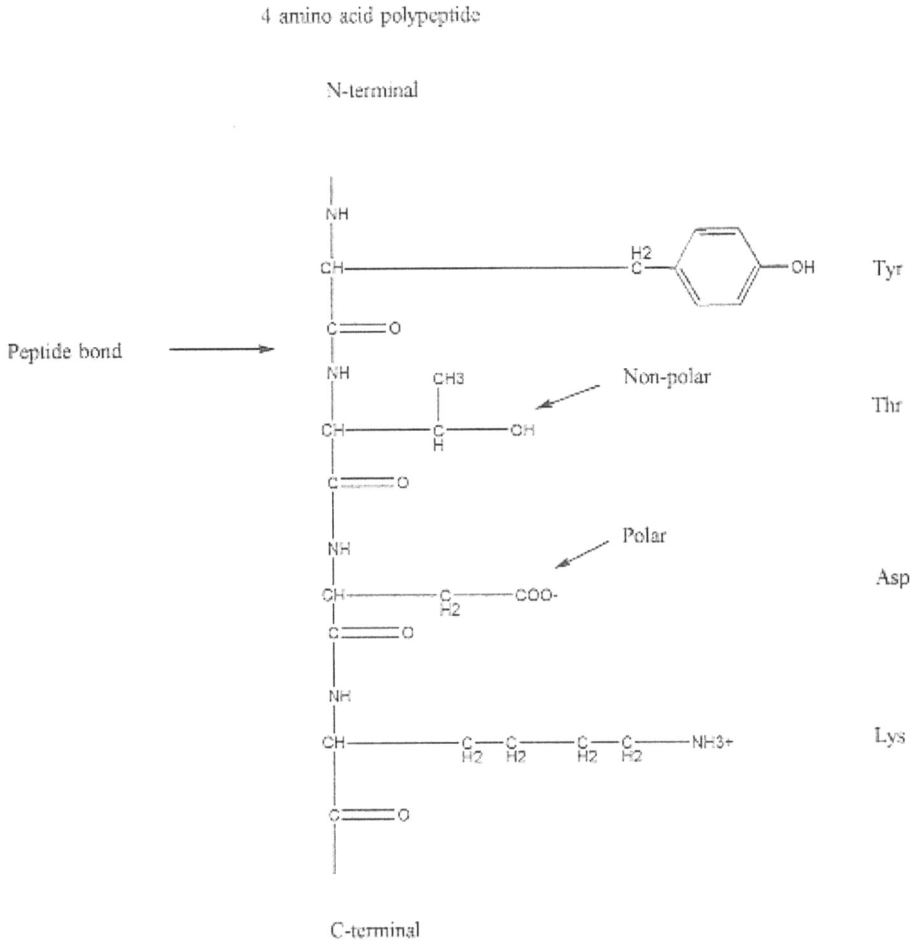

4 amino acid polypeptide

N-terminal

Peptide bond ⟶

Tyr

Non-polar

Thr

Polar

Asp

Lys

C-terminal

Solution: the alpha-helix
The alpha-helix is the major transport protein in eucaryotes (cell with nucleus).

Now we will take a closer look at the alpha-helix (35.):

Alpha-helix: Space-filling model (van de Waals radii)
The turns of the molecule are to the left=counter-clockwise.
The plane of the peptide bonds run parallel to the axis

O + N form H-bonds: C2HN: + C2O : : - - > C2N-H - - - O-C2

*6J Alpha-Helix Protein

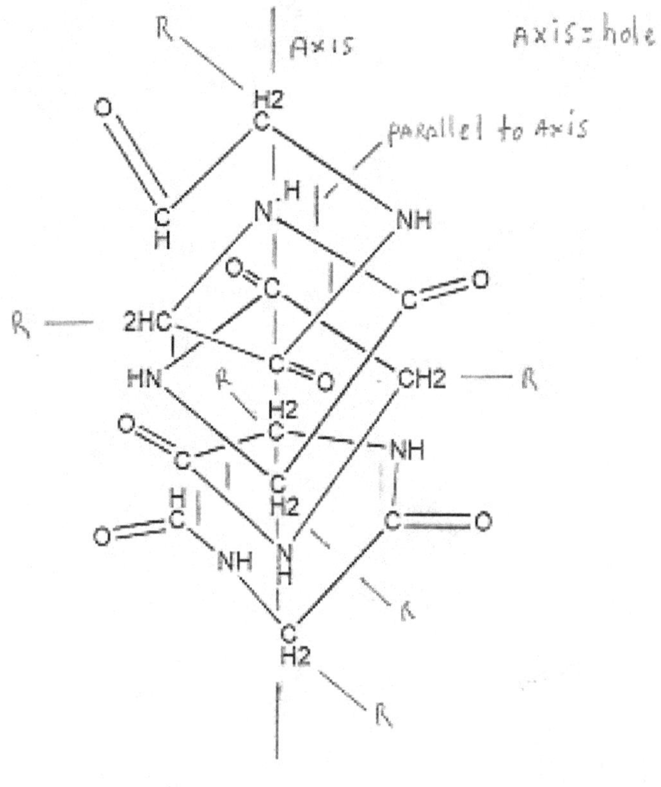

In the diagram, can you draw H-bonds (dotted lines) to each atom in the Alpha-helix?

A look down the hole (*36*):

Alpha-helix: Ball and Stick model
Axis pore: Allows Water and Electrolytes to pass through membrane
Wall: van de Waals radii makes it solid (no holes)
R group points outside

Can you identify the hole in the above diagram?

*6K Alpha-Helix Hole Ball and Stick Model

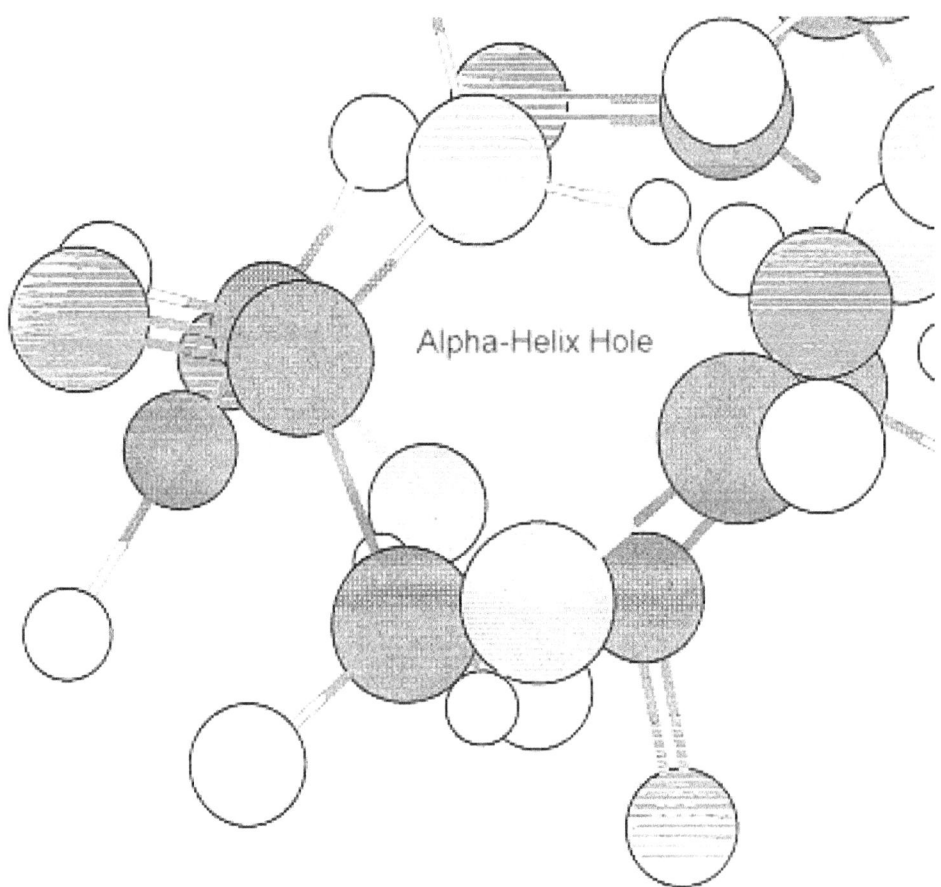

Alpha-Helix Hole

Beta-sheets are restricted to bacterial, mitochondrial, and chloroplast outer membranes (*37*).

Beta-sheet formed from Beta-strands: The shape of a barrel.

*6L Beta-Sheet made from Beta-Strands

Beta-Sheet

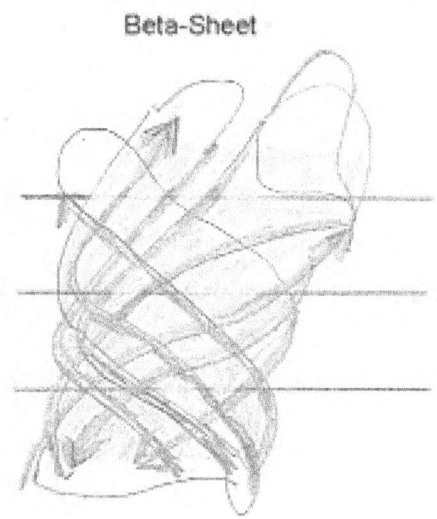

Summary: Proteins in the membrane are versatile in function.

The **alpha-helix** motif is the main secondary protein structure providing water and electrolyte transport via the membrane.

Water and electrolytes are not free to go and do whatever they please.

The lipid bi-layer isolates the cell and cell compartments from free passage of water and electrolytes.

The proteins allow entry and control of water and electrolytes.

Chapter 7

Membrane Transport

When one looks at the difference between Extracellular and Intracellular fluid compartments, you see the following: Extracellular Fluid contains large amount of **Na+** and Cl-. Intracellular Fluid contains large amount of **K+**, PO4-2 and Proteins. This is largely due to the protein function of the membrane; these proteins are called **transport proteins.**

*7A Na+ and K+ ECF and ICF

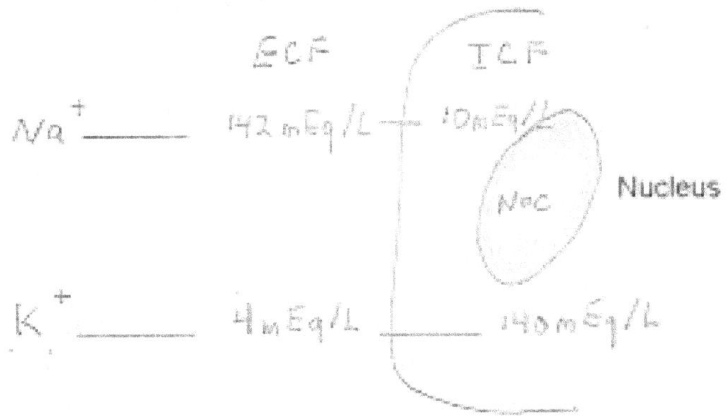

Note the closeness in amount of each ion between the compartments. There are 2 prevailing principles that you need to keep in mind. 1. **Osmotic balance**. The # of particles in a given volume of fluid need to balance in order to prevent the movement of H2O between the compartments, eg. Brain swelling after trauma. 2. **Charge balance** on each side of the membrane needs to be maintained. Special tissue elicit a charge imbalance to create a biological action of importance, eg. Nerve conduction.

These ions are in an aqueous state. So H2O can move according to osmosis.

Both fluid compartments have much more constituents than depicted [*37*].

The bilayer is permeable to **H2O**, small uncharged polar, and non-polar molecules.

Ions are in a hydrated state making them too large for passage.

Na+ - - - - OH2   H-bond.

There are 2 main types of transport proteins [*38*]:

> 1. **Channel** proteins: containing pores
>
>> Alpha-helix
>
> 2. **Carrier** proteins: bind to solutes.

These operate by 2 main physical mechanisms:

> 1. **Diffusion**:
>> A. **Simple:** via channel protein, kinetic energy of motion
>> B. **Facilitated** Diffusion: via carrier protein
> 2. **Active** Transport: via carrier protein, need ATP for energy.

*7B Protein Transporters

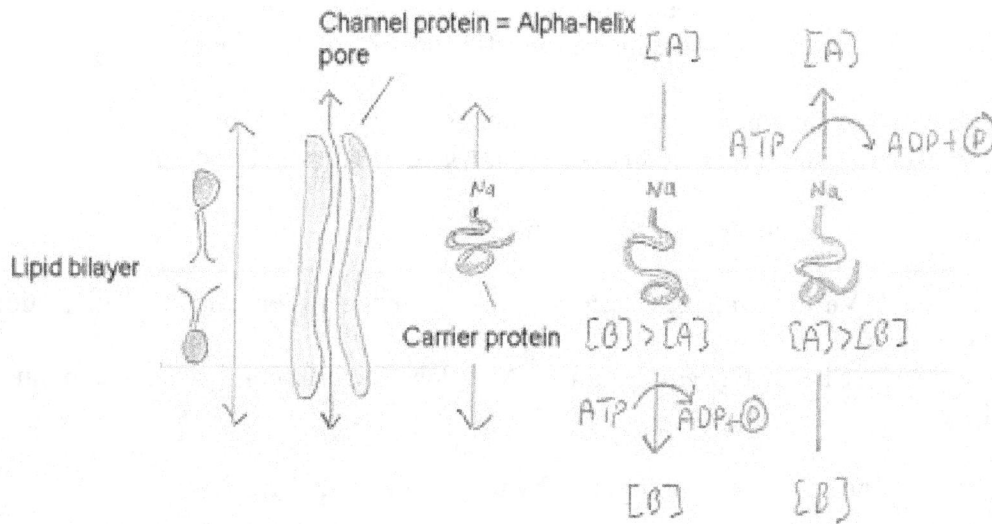

The first double-headed arrow on the Left depicts bidirectional nutrient crossing of the bilayer by **simple diffusion**. This is between the ECF and ICF compartments. The force driving the motion is the kinetic energy of motion (Heat)

The 2nd from the left is the **channel protein** which contains a pore (hole created in the membrane). The Alpha-helix is ideal for this mechanism. It too is bidirectional and work is done also by the kinetic energy of motion. The nutrient contains heat which creates motion and therefore crosses the membrane. It is like a machine gun firing bullets at a wall with a hole in it.

The 3rd one is the **facilitated** type which means that the nutrient is carried across the membrane by the protein itself. This is also bidirectional. This could occur with the "Flip-Flop" mechanism described earlier, simple protein diffusion across the membrane, or the **diffusion** along the surface of the protein by the nutrient with the protein acting as a bridge across the membrane. The protein undergoes conformational change (change in shape to move the nutrient through a given distance (the length of the membrane).

The last 2 are done by carrier protein, except this time, energy is consumed in the process. The nutrient is driven against a concentration gradient. The higher concentration on one side prevents the movement by the nutrient under the kinetic energy of motion.

To overcome this barrier in energy, energy is derived from the splitting of ATP to ADP + PO4(-2) + energy. The energy is used possibly to change the shape of the protein with the end result of transporting the nutrient against a higher concentration. This process is called **Active Transport**. Obviously, this process is unidirectional.

The protein channel possess **selective permeability** for the following reasons [38]:

     1. nature of the **channel**: diameter, shape and electric charge
     2. **chemical bonds** inside along the surface of the channel.

The **Na+** channel is .3x.5 nm in diameter and (-) charged. [40].
The **K+** channel is .3 x .3 nm and no charge [41].

These channels are also gated by 2 mechanisms [39]:

     1. **Voltage** gating: via charged dipoles in the protein channel, eg. Action potential: nerve cell
     2. Chemical (**ligand**) gating: Acetylcholine channel, eg. nerve and muscle transmission.

The gated channels operate by the "all or none" principle.

*7C Gated Proteins

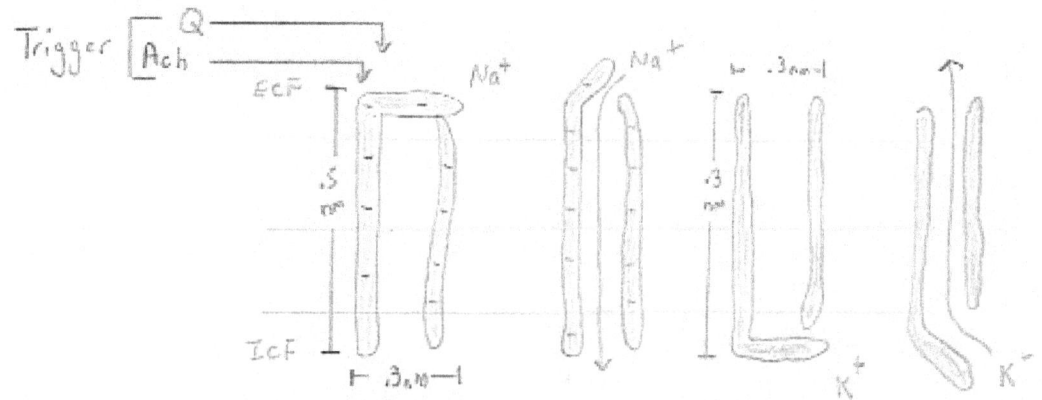

The channel protein can have a gate mechanism. The trigger for opening the gate can be electrical or chemical. An electrical trigger is an electrical potential created by a charge difference in the membrane itself which in turn causes a conformational change in the channel protein which opens the gate for nutrient passage. The is facilitated because regions of the

protein have dipoles which assist in the conformational change. The Na+ channel protein also has a net (-) charge which attracts the Na+ ion down the channel. The chemical trigger is called a ligand because in bonds/interacts with the protein to trip the gating process. The gated process is unidirectional. It is also "all or nothing" in principle. This means it is either open or closed, no in between state.

Osmosis [42]:Simple Diffusion, net movement, of **H2O** caused by a difference in concentration of water. Osmosis stops when the pressure equilibrate.

*7D Osmosis

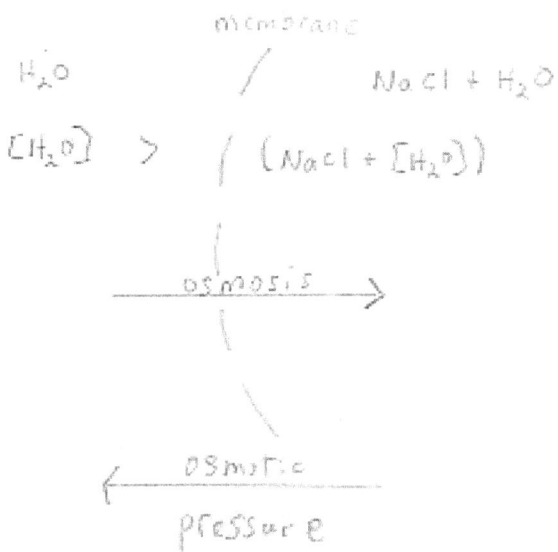

Physical barrier: Selective Permeable membrane
Force: Kinetic energy: $k=(m*v^2)/2$ m=mass v=velocity

Osmotic pressure: opposes osmosis. The amount of pressure to stop osmosis.

Osmotic pressure:

The # of particles/volume = molar concentration

1 osmole=1 gram molecular weight of osmotically active solute.

Osmolality: 1 osmole /kg of water

Normal ECF and ICF: 300 mOsm/kg [43].

{1 mOsm/L = 19.3 mmHg of pressure
19.3 mmHg * 300 mOsm/kg = 5790 mmHg
5790 mmHg * .93 = 5500 mmHg measured
(Molecular interaction: Attraction Na+ to Cl-)}[44].

Osmolarity: Osm/L of solution.
Difference in Osmolarity v. Osmolality <1%
because of dilute solutions of the body [45].

Several comments for clarification:

Kinetic energy is the energy of motion. This is occurring at body temperature of 37 degrees C [46]. Thermal energy is energy of heat. A higher temperature transfers the energy to kinetic energy of the particle.

Heat - - -> E k

The faster the motion, the more of the # of particles striking the membrane.

E k - - -> Force

Pressure is related to Force per area.

Pressure = Force/Area

Molar concentration is related to Avogadro's constant (N A). The # of particle in a mole of substance.

Mol * 6*10^23 = #

Concentration is term used in describing volume in Kg or L.

Molecular Weight is the amount of grams of a substance in 1 mole.

1 mol = M.W.

eg. Glucose's M.W.=180 gs. = 1 mol of glucose

Glucose: 6 C   6*12 (atomic mass unit) = 72 g
12 H   12*1   = 12 g
6 O   6*16   = 96 g
Total: 180 g M.W.= 1 mol

Osmole is 1 mole of osmotically active substance.

Glucose does not ionize, therefore 1 mol of Glucose=1 Osmole of glucose.
Ionize means to dissociate into ions. eg. NaCl (1 particle)- - -> Na+ + Cl- (2 particles)

Active Transport: From low to high concentration:

2 types based on energy requirement:

1. Primary: ATP hydrolysis
2. Secondary: [ ] gradient

Both carrier proteins

*7E Na+ - K+ ATPase Enzyme

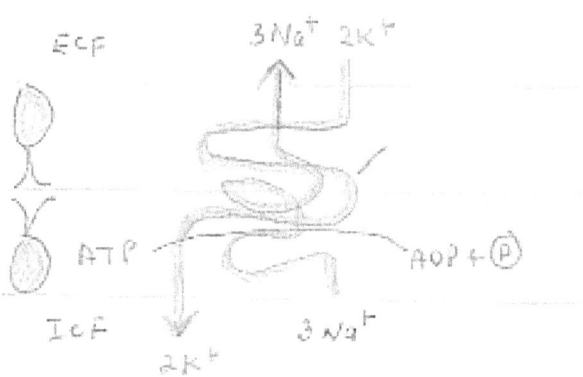

N+-K+ pump[*47*]:
2 subunits: α: 100,000 daltons:
Receptor + ATPase
β: 50,000 daltons
Membrane anchor
3 Na+ out
2 K+ in
High energy cost
Reversible

Control of cell volume: **H2O** management via Na+ (osmosis)
Net +1 Q outside of membrane: 3-2=1
Energetics: E=1400 * log C1/C2 (cal./osmol)

Secondary Active Transport:

Co-transport: same direction
Na+   +   Glucose
Na+   +   Amino Acids

Counter-transport: Opposite direction
Na+   +   Ca+2
Na+   +   H+

Clarification: Dalton= 1/12 mass of C12   (Atomic mass unit)
Log (C1/C2) [ ] magnitude   Log= base 10
When C1>C2 – > E is (+) = requires energy input
When C2>C1 –> E is (-) = spontaneous reaction
For ion transport, Energetics equation transforms to:
$E = 1400*\log C2/C1 + z*F*\Delta\psi$

{z=charge on molecule [48].
F (faraday)=23,062 cal/mol*V
$\Delta\psi$ =membrane potential in Volts
1400 = calories
2.0303 *R*T
R=gas constant=1.987 cal/mol*K
T=absolute temperature= 273 C + 25 degrees}

Tissue Transport [49]:

eg. Gastrointestinal Tract Epithelium

Small Intestine

Luminal side: Simple or Facilitated Diffusion

Basement Membrane side: Active Transport + Osmosis

Cell Junctions: Same as B.M..

*7F Tissue Transport

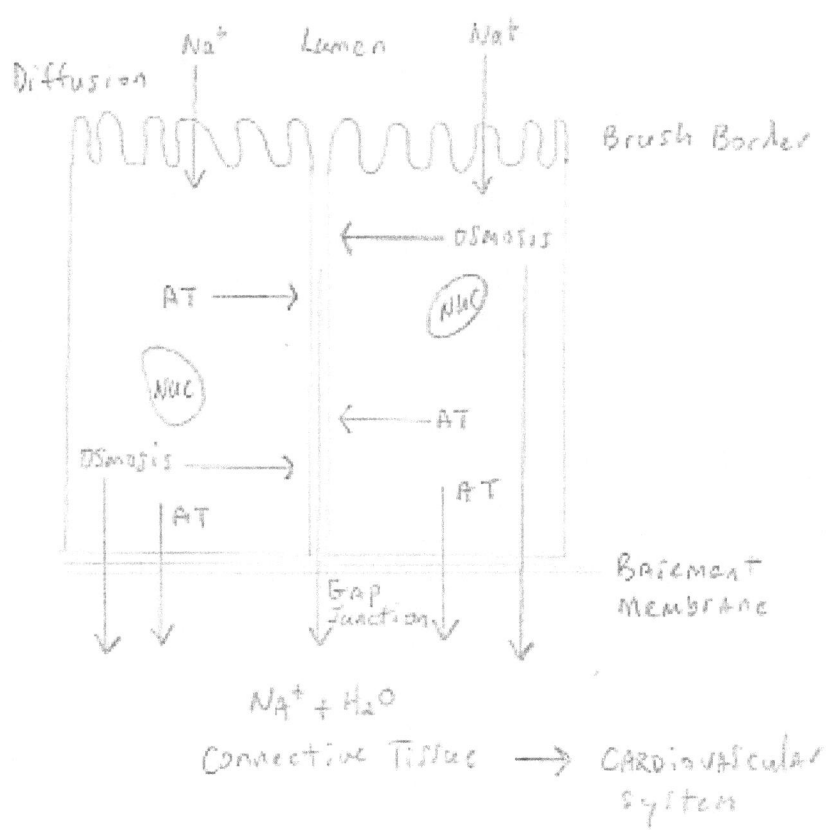

Lumen=inside the bowel.
Brush border=microvilli surface of epithelial cell, surface of absorption.
AT=Active transport
Gap Junction=space between adjacent cells.
Basement membrane=border between epithelial cell and connective tissue.

Na+ is actively transported out of the cell and H2O follows by osmosis.

Lumen –> Epithelial cell membrane –> ICF –> Connective Tissue (ECF)

Summary: All the transport mechanisms are involved in **Na+** and **K+** metabolism. From the point of ingestion of these micro-nutrients to their final level of physiologic action. **H2O** follows Na+ by osmosis in order to maintain cell volume.

Chapter 8

Anatomy of Water

      Body fluids are controlled by a strict homeostatic system. Both intake and output are highly variable. To keep control of this, the body utilizes the kidney and thirst mechanism to control the internal environment. We start with the anatomy of water, compartmentalization [50].

      *8A Daily I and O

Daily Intake and Output of H_2O (ml/day)		
	NORMAL	Exercise
Intake		
Ingested	2100	x
Metabolism	200	200
Total	2300	y
Output		
Skin	350	350
Lungs	350	650
Sweat	100	5000
Feces	100	100
Urine	1400	500
Total	2300	6600

      We see a strict control of the intake and output. Under normal conditions, the totals are equal. This is a balanced condition in which the body maintains with its homeostatic system.

Fluid from food is because of the combustion of, for example carbohydrates, leading to the products of CO2 and **H20.** Insensible loses are loses which are not felt by the person.

      Look at the derangement created by prolonged, heavy exercise. The loses from the lung and **sweat** are tremendous. **50 X** (5000/100) normal in sweat and 1.86 X (650/350) normal in the lung leading to a total magnitude of lost of 2.87 X (6600/2300). The reason for this is the dissipation of **heat** generated by the working muscles. Heat overload can by deadly.

By the way, 5000 ml = 5 Quarts of fluid =1.25 gallon (5/4)

Air entering the lung holds less moisture then the air in the lung. So water is transferred to the expelled air. This is worst in colder weather and is lost at a faster rate during exercise.

The urine output drops during exercise to conserve water due to the loses in sweat. There is a lower limit because of the need to rid the body of nitrogenous waste which is toxic.

X means we can control our intake through a complicated neurophysiologic mechanism
and therefore change the outcome, total Y. We can do this ourselves provided we are conscious.

Water is compartmentalized [*51*]:

*8B Water Compartments

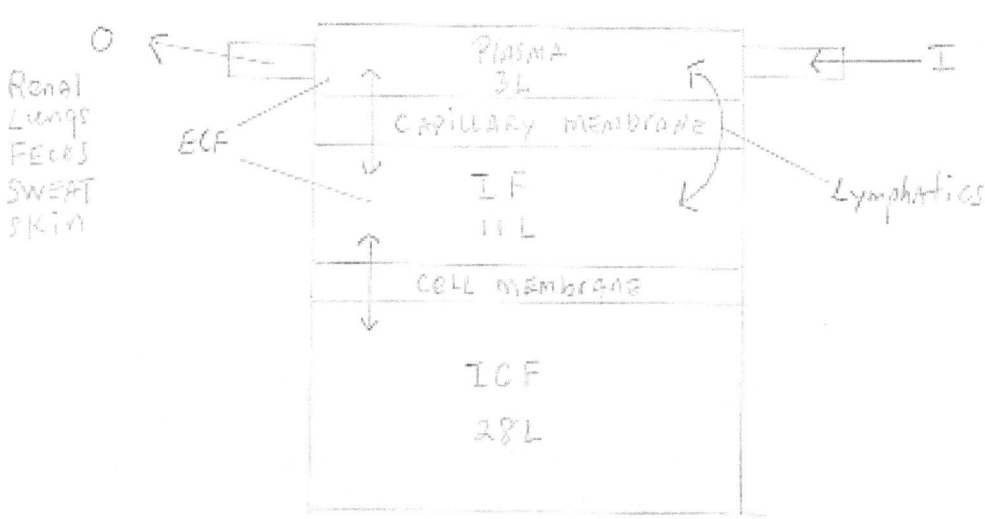

Total Body Water 42L (60% of body weight, 70 kg. Average)
Intracellular 28L (40%) (Inside the cell)

Extracellular 14L (20%) (Outside the cell)
Interstitial 11L (15.7%) (Protein filtrate, contains no protein) (Outside and between cells)
   Plasma 3L (4.29%) (60% of blood) (In veins, arteries, and lympahtics)

Transcellular 1-2 L (2.86%) (Special compartments eg. Synovial)

Blood 5L (7% of body weight)
RBC 2L (2.86%) (40% of blood)

   TBW(Total Body Water)=E.C.F. (Extracellular Fluid) + I.C.F. (Intracellular Fluid).

   Most of the body is made of water. When divided, most lies in the cell compartment. This is where all the vital life processes take place. The next largest are is between the cells. It is of interest to note, that the water compartment that is involved in transport and maintenance of the cell is the next to smallest (Plasma, 4.29%). This is in keeping with the concept of conservation of matter and energy. Also, this can become a big concern with small disturbances in this area, eg. Septicemia, dehydration (sweat). The body relies heavily on the ability to return to homeostasis.

   Of the dissolved constituents in water, are they the same in all compartments? We need to discuss the chemical anatomy of water. Water is rarely in a pure state. It usually has dissolved
and undissolved compounds associated with it.

E.C.F. IF (Interstitial Fluid)   I.C.F. [52]

Protein(Plasma)   **Protein**
Na+   **K+**
CL-   PO4-
HCO3-   SO4-
Ca2+   Mg+2

E.C.F.= I.F.   The horizontal line in diagram is the 0 line.

*8C Na+ and K+ ECF and ICF

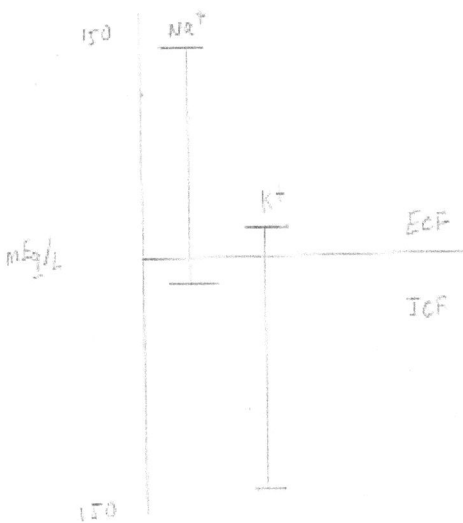

Looking at the chemical constituents of water by a osmolar perspective. (mOsm/L H2O) [53]:

	E.C.F.	I..F.	I.C.F.
Protein	1.2	.2	4
Na+	142	139	14
K+	4.2	4.0	140
Total mOsm/L	301.8	300.8	301.2
c mOsm/L	282	281	281
Total osmotic pressure	5443	5423	5423

37 C   degrees
c=corrected
Osmotic Pressure in mmHg

I.C.F protein is 3.33 X (4/1.2) Plasma protein.

The effects of proteins on ions will be further explained.

Total and corrected osmolarity and osmotic pressure will be discussed shortly.

Chapter 9

The Donnan Effect

Because of the high concentration of protein in the plasma, a Donnan effect occurs between the E.C.F. and I.F. See figure 1. The protein having a negative charge creates an electrical force with other charged compounds or ions. The physical nature of the Donnan effect is found in Coulomb's Law: **Like** charges **repel** and Unlike charges attract. See figure 2. The attraction is non-specific: both Na+ and K+.

*9A The Donnan Effect

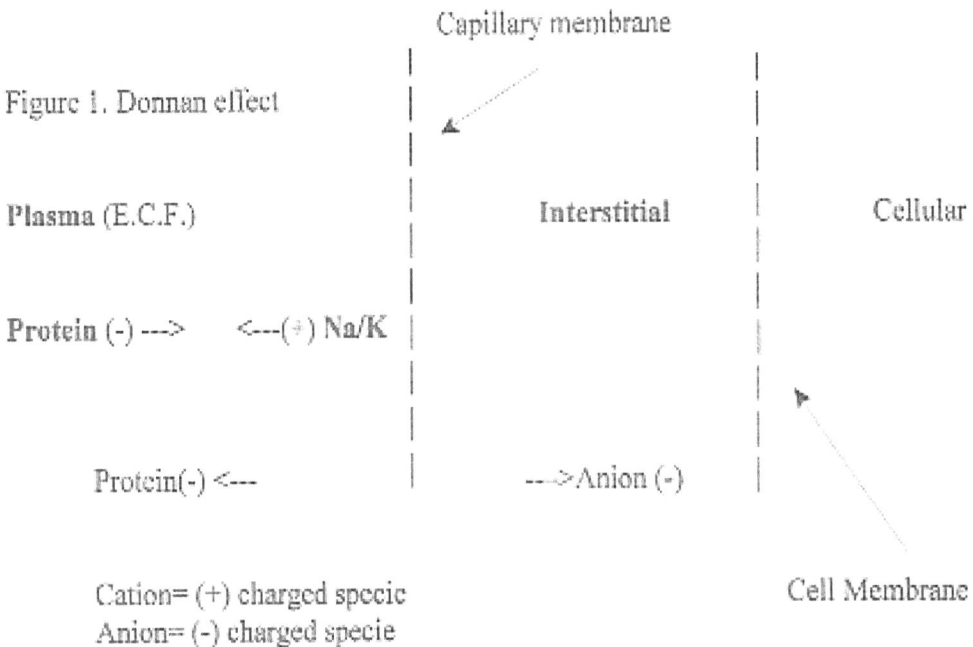

Figure 1. Donnan effect

The barrier between the E.C. and Interstitial department is the highly permeable **capillary membrane** which is permeable to electrolytes, water **not proteins.** Between the interstitial and cell is the **cell membrane** highly permeable to water **not electrolytes or proteins.** This design is capable of controlling ionic forces and osmosis. Separation of cations sets up a potential energy differential and along with protein creating a osmotic pressure differential. The differential is force by nature. Water will tend to enter the circulation and cell compartments. A net charge will set up between the interstitial and cell compartments.

*9B Coulomb's Law

Figure 2. Coulomb's Law

The formula : $F = (q1*q2)/r^2$

q1 + q2=charges
F=force
r=radius

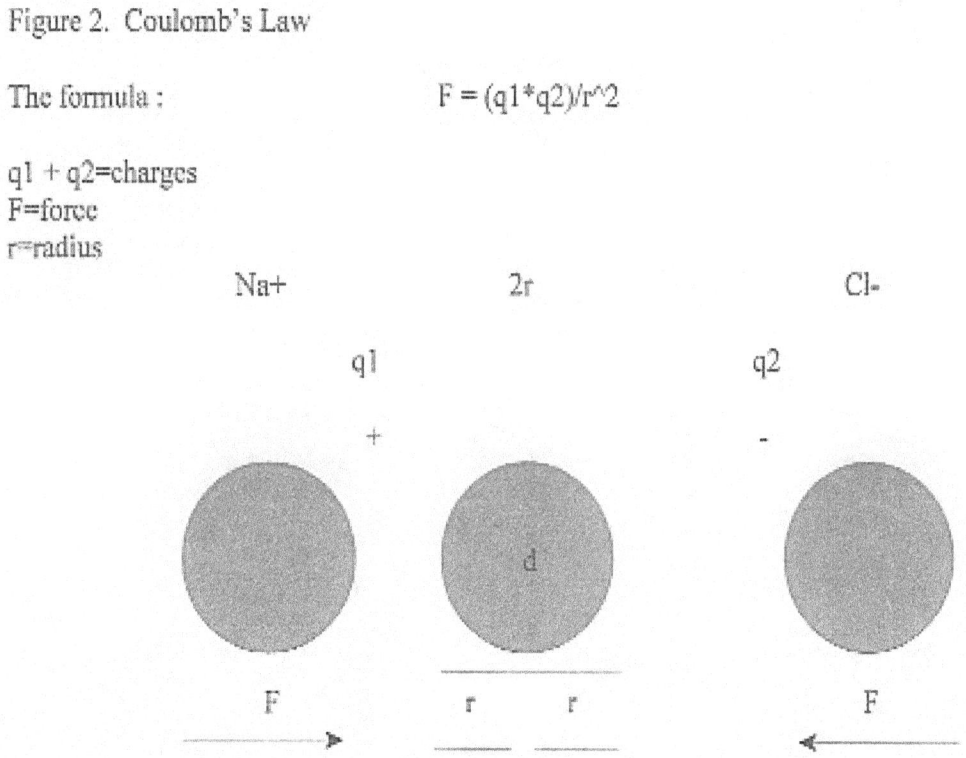

This is the same law that predicts the large **dielectric constant** of water. This enables ionic compounds to **dissolve** in water because water reduces the attraction force that holds the ionic species together. **Polar water** dissolves ionic compounds. This is the process of **hydration.** The **H-bond** plays a pivotal role in this process.

*9C Force v. q Graph

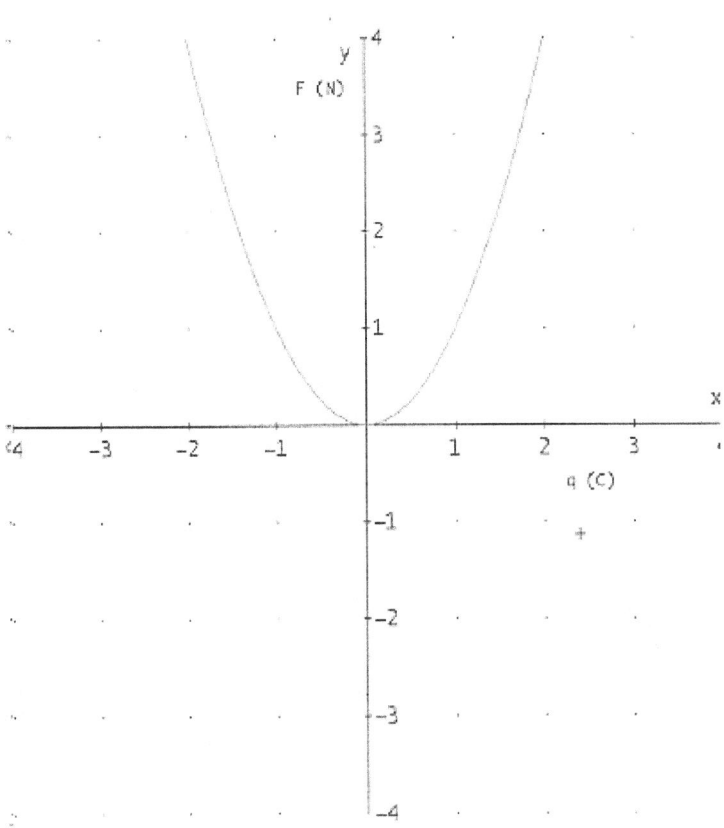

Newton: kg*m/s^2
Coulomb: Ampere*s
Ampere: q/s

Y-axis divides the graph into to mirror images: symmetry along the y-axis.
If q1*q2= (+) q^2 then the force is a type of Repulsion.
If q1*q2= (-) q^2 then the force is a type of Attraction.

*9D Force v. r Graph

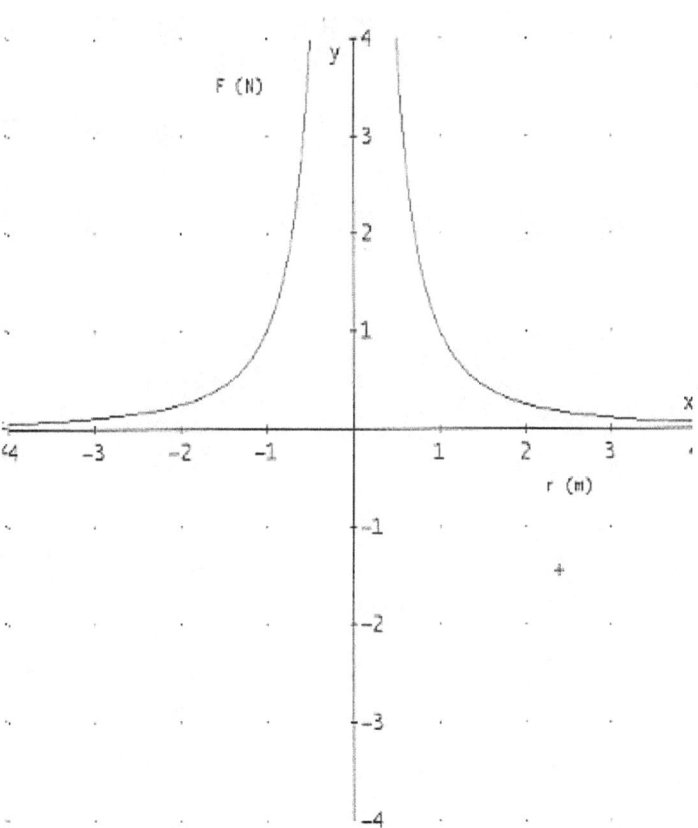

Radius is the distance of separation between the charges (q).
As the radius gets larger the F gets smaller.
Maximum F is close to <1 r.
Symmetry along the y-axis.

NaCl + H2O ----> Na+ + Cl- + H2O (NaCl dissolved in water, state of Hydration)

*9E Salt Hydration

Hydration

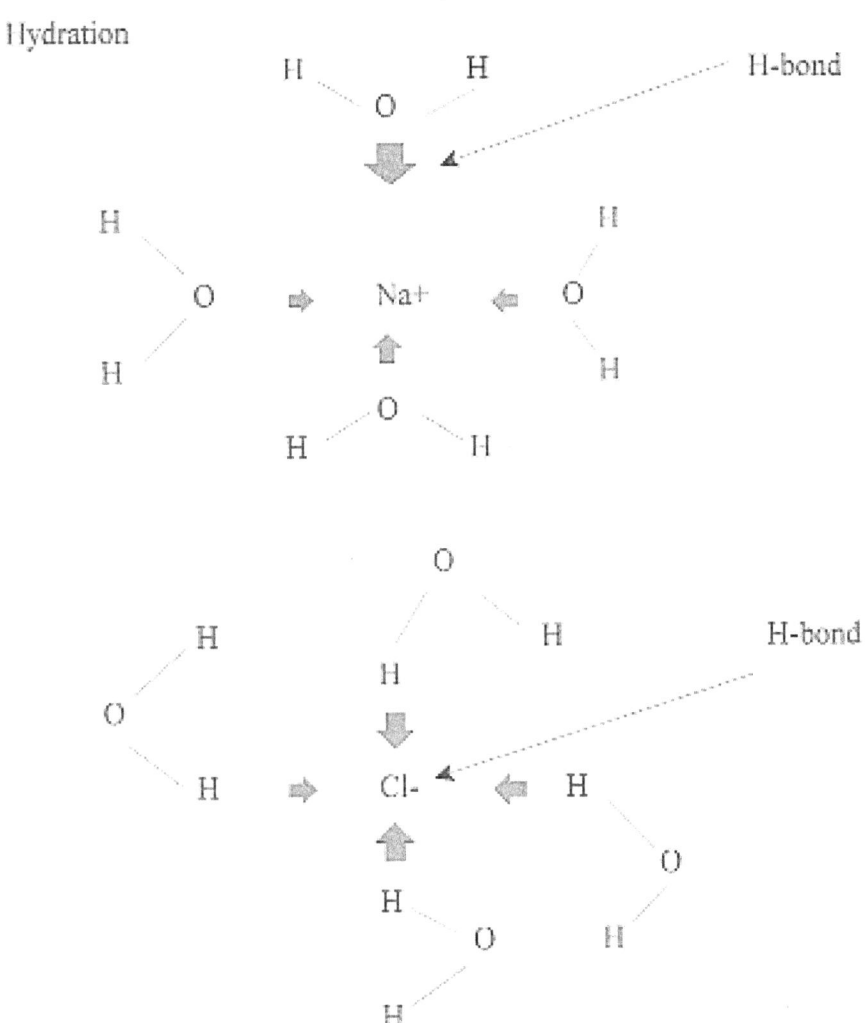

2 other properties of water:

[H2O]: High molar concentration
Most abundant compound in the body.
Large source of H+.

Hydration
Metabolism (platform for chemical reactions)

[$H_2O \longrightarrow H+ + OH-$]: Dissociation constant is very small.
Controls the H+ radical.
Controls the **pH** of the system.
Stabilizes **cellular** and metabolic (**enzyme**) environment.

Chapter 10

Starling Forces

Starling forces are the drivers for fluid regulation between compartments. These are:
1. **Osmotic** pressure
2. **Hydrostatic** pressure
    Hydro: water (solvent)
        static: not changing
    The weight of water (kg) or Volume (m^3) **standing** against an area of surface (m^2).
    kg: kilogram=1000 grams
    m: meter=1000 millimeters
    milli: 1/1000 of meter
    pressure [Pa: kg/(m*s^2)]=F/A (Force/Area)
        Pa: Pascal
        F: Newton (N) : kg*m/s^2 = m*a
        m: mass (kg)
        a: acceleration:(m/s^2) = change in velocity (m/s)
    Source: Blood Pressure (BP)
    C.O.(Cardiac Output: L/minute) = S.V.(Stroke Volume: L/beat) *
H.R.(Heart Rate: beats/minute)
        Ohm's Law: B.P.(Blood Pressure: mmHg) = C.O.(L/min) * PR
(mmHg/ml/sec)
        mmHg: the physical application of the mercury manometer.
        P.R. (Peripheral Resistance): The impediment to blood flow
exerted by the endothelial (inside) lining of the blood vessel.

Hydrostatic pressure is a physical concept. To understand the physics of the macro circulation, dimensional analysis was used on the above equations to fully understand how these equations are derived. Dimensional analysis is using the equations in unit form; paying attention to the units. After studying the equations, hopefully you will understand how they are applied in medicine (biology). The material above was an attempt to demonstrate the origin of the Hydrostatic pressure. This pressure though very diminished at the level of the capillary (micro circulation) plays a key role is the transfer of fluids and ions to the cell environment. So the macro circulation is part of the homeostatic system acting a a blood conduit which delivers necessary nutrients to the cell. To study further the mercury manometer, I refer you to the text: Physical Chemistry by Peter Atkins, Freeman Co., 6th Edition, page 14.

As a historical note, the manometer was invented by Torricelli (an Italian physicist and mathematician), a student of Galileo. The N (Newton) is the unit of force named after Sir Isaac Newton. The Pa (Pascal) is the unit of pressure named after Blaise Pascal (1623-1662), French mathematician and philosopher.

From macro to micro:

Capillary wall [*54*]:

*10A Intercellular Cleft

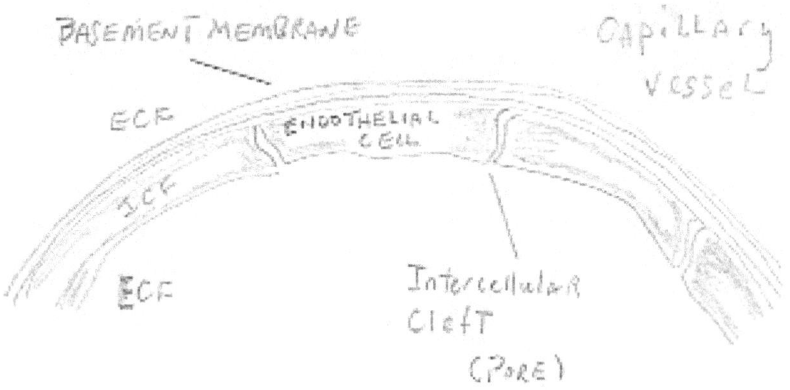

Pore=intercellular cleft (nutrient passageway)

Protein not permitted

Permeability (Size) [55]:

*10B Pore Permeability

Relative Permeability of Pores

Substance	molecular Weight	Permeability
Water	18	1.0
NaCl	28	0.96
KCl	36	0.96

H2O : 1
Na+ : .96
K+ : .96

Protein: .001

Starling's Forces in equation form:

Determines the movement of fluid through the capillary membrane.

$$NFP = Pc - Pif - OPp + OPif$$

Mean (mmHg)

	Mean (mmHg)
NFP : net filtration pressure (out of capillary)	.3
Pc : capillary pressure (out)	17.3
Pif : interstitial fluid pressure (in)	-3
OPp : plasma colloid osmotic pressure (in)	28
OPif : interstitial fluid colloid osmotic pressure (out)	8

colloid : description of protein in dissolved state
oncotic : osmotic pressure at the capillary membrane, not cell membrane.
Pif : is (-) due to lymphatic drainage from interstitial fluid compartment
     In equation, Pif is - -Pif –> + Pif
OPif : Protein leakage through capillary membrane
Overall results : 3 forces outward + 1 force inward = net outward (homeostatic state).
              .3 mmHg is the result of the lymphatics.

If NFP is (+), net diffusion out of capillary.
If NFP is (-), net diffusion into capillary.

     Rate of capillary fluid filtration:

     Filtration = Kf * NFP

Kf : capillary filtration coefficient
     Capacity of capillary membranes to filter water for a given NFP (ml/min*mmHg)
     A fraction of 1.

     Summary: Hydrostatic and osmotic pressures together determine the net flow of
H_2O, Na^+ and K^+ from E.C.F. to Interstitium to Cell compartments. It is a bi-directional
flow. It can reverse itself.

Chapter 11

Osmosis

Osmotic pressure: The pressure which stops osmosis.
Source: proteins + ions (solute)
Diffusion: [High]–>[Low] : Thermal (kinetic: motion) energy
Osmosis is the diffusion of water across a selectively permeable membrane.

*11A Diffusion of Water : The Concentration Gradient

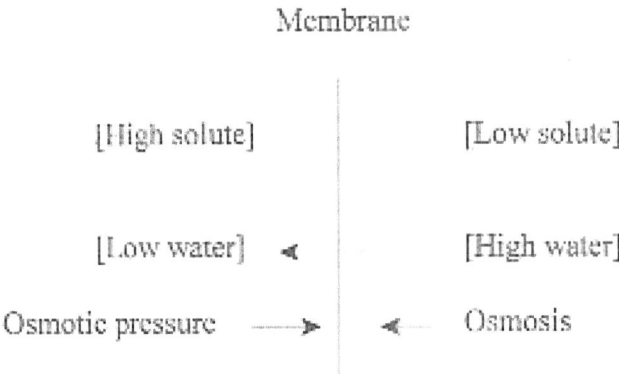

Diffusion: passive event
      mainly driven by thermal motion (kinetic energy) of water molecule.
    Osmoles is the total # of particles in solution.
    Osmolarity is the number of osmoles/L.

According to van't Hoff's law:

$$OP = CRT \ [56]$$

OP = osmotic pressure=mmHg
C= osmolarity=mOsm/L
R=gas constant=62.3 mmHg (L K mol)
T=Kevin=(273 +37) degrees

So at 1 osm/(L=1000 ml) * 62.3 mmHg*310 degrees = 19,300 mmHg
19300 mmHg/mOsm/L=19.3 mm Hg

In magnitude: 1000* 62.3^(-3)*310=19300
19300/1000=19.3^(-3)

1 mOsm concentration gradient across a membrane=19.3 mm Hg

Example of calculation: 0.9 Normal Saline
.9%=.9 gms./100 ml = 9 gm/L
Molecular Weight= 58.5 gm/mol=22.99+35.45 (The Periodic Table, amu (atomic mass unit))

9/58.5=.154 mol/L
@ molecule=2 osmoles
osmolarity=2*.154=.308 osm/L*1000=308 mOsm/L
osmotic pressure=308 mOsm/L*19.3 mmHg/mOsm/L=5944 mmHg
308*.93=286
.93=osmotic coefficient (interionic attraction) 2–>1 osmole

Let's look at the fluid compartments:

Osmolarity [57]

	Plasma	Int.	I.C.
	Na (142)	Na (139)	K (140)
	Cl (108)	Cl (108)	other (161.2)
total	301.8	300.8	301.2
c mOsm/L	282.0	281.0	281.0
mmHg	5443	5423	5423

EC: Na + Cl = 250/301.8=.828*100=82.8%
IC: K=140/301.2=.465*100=46.5%
c=.93 eg. 301.8*.93=282
Plasma: 1 mOsm/L larger because of proteins exerting 20 mmHg   282-281=1
5443-5423=20 mmHg
5443=282*19.3

In the example of calculation, we calculated the osmolarity and osmotic pressure

of a solution, namely .9 Normal Saline. We found that it was very close to body osmolarity and pressure making it ideal for fluid replacement. Because of the similar characteristics, .9 Normal Saline will not disturb the water balance in the body to any great degree. This fluid is used
primarily in medicine.

Let's look at 3 common oral hydration solutions; do the same calculation to 1 of them and compare the results. We will start with the Nutrition Facts of each product.

Nutrition Facts [58]

	Ocean Spray Cranapple	Gatorade	Powerade
Serving size:	8 Fl. oz. (1 cup)	8	8
Calories	140	50	60
Total Fat	0 g	0	0
Na	80 mg	110	55
K	15 mg	30	30
Total Carbohydrate	35 g sugar	14	17
Protein	0	0	0
	Vit. C		Vit. B3,6,12

Looking at the above solutions, we find that they differ very little with each other. The calories of all 3 solutions come strictly from the sugar. The calorie formula is:

CHO. g.* 4 cal/g.=total calories.

Notice the amount of the electrolytes, the magnitude in milli.=1/1000 g. The mole content is very small. This may predict a very small osmole content; thus making these solutions have small osmolarity and pressure. The juice made from concentrate contains corn or beet sugar. The sugar could be glucose or sucrose, nature's sugar (also fructose), which is comely used in food processing. Usually added sugar is in the form of sucrose, a cane sugar. Sucrose is made of glucose and fructose. The fructose and glucose are absorbed from the intestinal track by facilitated diffusion, protein carrier mediated. The sucrose requires digestion by the sucrase enzyme and then is absorbed. Glucose is rapidly metabolized in the liver, but fructose requires conversion to glucose first. It is a known fact that a sugar load to the stomach may cause stomach retention; prolonging the clearance of the sugar from the stomach. This could lead to vomiting. So the timing of the load is important for clearance. The usual clearance time is approximately 30min to 2 hours. Pre-game hydration should recognize this time limit. Our program at Neptune required pre-game hydration 2.5 hours prior to start time. Once ingested, the fluid reaches all points in the body is about 30 minutes [59]. It has also been shown that a

water diuresis starts in 30 minutes and continues for 2 hours, during which the urine output increases
8X normal [60]. Therefore, we had our players void 30 minutes prior to start time to empty the bladder.

Let us return to the example calculation, but this time we will use Cranapple:

Molecular weight (Periodic Table):   Glucose (assumed): 180 g/mole
C: 6 *12 g/mole=72 g/mole
H: 12*1 g/mole=12 g/mole
O: 6*16 g/mole= 96 g/mole
Na+ :   23 g/mole
K+:   39 g/mole

Calories: 35 g * 4 cal./g= 140 cal.

8 oz. * 30 ml/oz.=248 ml* 1L/1000ml=.248 L

Na+: 80 mg*1g/1000 mg=0.08 g/.248 L=.3226 g/L * 1 mole/23 g=.014 mol/L*1

osmole/mole=.014 osmole/L * 1000 mOsm/osm= 14 mOsm/L/serving = osmolarity

OP = 14 mOsm/L*19.3 mmHg/mOsm/L=270.2 mmHg = osmotic pressure

K+: 15mg=.015g/.248 L=.0605 g/L * 1mole/39 g=.00155 mole/L * 1 osmole/mole=.00155 osmole/L = 1.55 mOsm/L

OP=1.55 * 19.3=29.9 mmHg

Glucose: 35g/.248 L=141g/L/180 = .783 mole/L * 1=.783 osm/L=783 mOsm/L

OP = 783*19.3= 15112 mmHg

	Osmolarity	OP
Na+	14	270.2
K+	1.55	29.9
Glucose	783	15112
Sum	798.6	Sum 15412

The osmolarity of Cranapple is greater than the body osmolarity, 798.6 > 282 (2.83X), making this solution hypertonic. The gluose is the majority compound. The osmotic pressure exerted is 15,412 (2.84X) which is greater than the body osmotic pressure, 5423. The overall immediate effect on water is to pull water out of the cell. This makes sense to supply the athlete water during the game to counter the hypertonic effect of Cranapple, with the kidney clearing the free water excess. Also during exercise, a later effect, the glucose will be metabolized to CO_2 + H_2O. Adding a small volume of water to the body.   This is an approximation to the true osmolarity for the following reasons:

1. There are more ingredients in the product other than the Nutrition Facts. These too have osmotic effects. The osmolar content is unknown. There is no weight stated about these compounds.

2. Sugar in food comes in many forms [61]:

Corn	Beet	Cane	High Fructose Corn Syrup
Glucose	Sucrose	Sucrose	Isoglucose =glucose + fructose=enzymatically treated glucose to create a certain % of fructose.
1 osm/mole	2	2	1

In our calculation, it is un-declared which sugar is used. If sucrose is used, there will be a multiplication factor of 2 in the calculation. If HFCS (fructose>glucose %) is used, there will be 1. One can determine the osmotic pressure of a solution by using the osmometer [62]. The main reason for using sport drinks for hydration is not the sugar content which provides some energy, but mainly for the replacement of Na+ and K+ which is lost during exercise.

*11B Sweat and Cranapple Effect

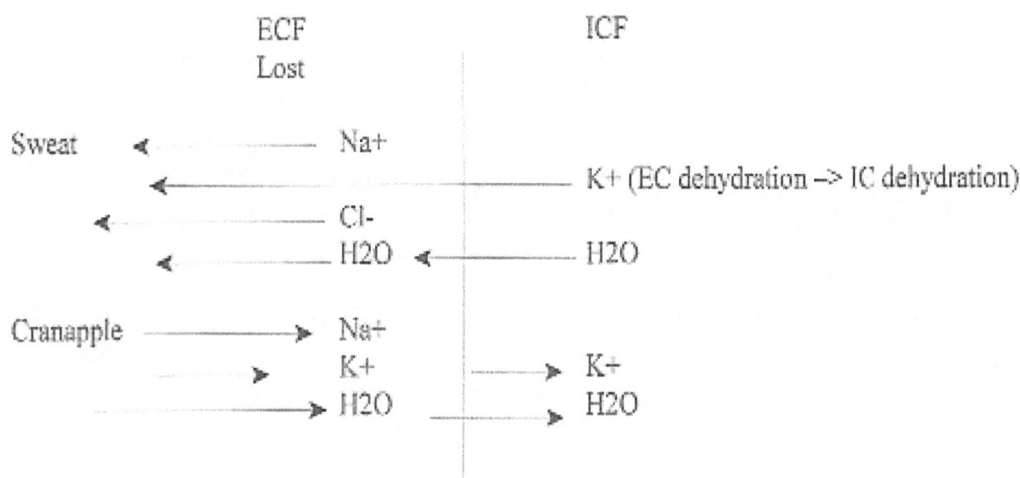

f you determine the osmotic pressure of a solution. You can calculate the total osmolarity of the solution with the following formula:

$$OP /19.3 = osmolarity$$

Now, we will discuss the concept of osmotic equilibrium. But you need to keep in mind the following idea: small changes in the EC can create big changes in the IC. Osmotic equilibrium occurs due to a physical force, the concentration gradient. It is automatic in nature. It is predictable and expected to occur when a change in solute is present. So one needs to consider 2 compounds during this equilibrium process: the solute=Na+ and the solvent=H2O. Also keep in mind that this equilibrium process can take seconds to minutes. Remember that it will take 30 minutes for the entire process to occur throughout the entire body after a drink. So as we proceed keep these 2 principles in mind 1. Water is going to shift between compartments and 2. Solutes are not.

*11C Na+ Load Effect

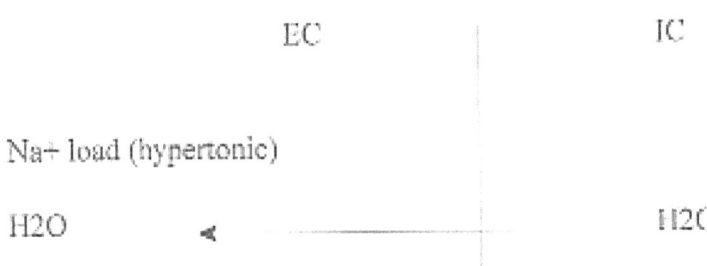

Summary of effects of a hypertonic solution to the EC.:
1. EC osmolarity increases. Na+ stays EC.
2. Water leaves IC
3. **ECV increases** (ECV change > volume load)
4. **ICV decreases**
5. **EC + IC osmolarity increases**

The EC osmolarity increases due to the Na+ load (addition of Na+ to the EC). The IC osmolarity increase due to the water lost.

We will do a calculation using Cranapple as the Na+ load. We have a 70 kg. athlete with an intial plasma osmolarity of 280 mOsm/L drinking 8 Liters (approximately 2 Gal.) of Cranapple. Remember the ECV=20% of body weight and ICV=40%.

Initial Conditions (No changes, at equilibrium prior to load)

	Volume (L)	Concentration (mOsm/L)	Total (mOsm)
EC	14	280	3,920
IC	28	280	7,840
Total	42	280	11,760

70 kg*.2=14
14*280=3920

Next, calculate the total mOsm added to EC in 8 L of Cranapple.

783 mOsm/L * 8 L = 6264 mOsm

Instantaneous Effect (No equilibrium)

	Volume (L)	Concentration (mOsm/L)	Total (mOsm)
EC	22	462.9	10,184
IC	28	280	7,840
Total	50	No equilibrium	18,024

10,184 mOsm/22 L=462.9 mOsm/L

At Equilibrium

	Volume (L)	Concentration (mOsm/L)	Total (mOsm)
EC	28.25	360.48	10,184
IC	21.74	360.48	7,840
Total	50	360.48	18,024

18,024 mOsm/50 L= 360.48 mOsm/L
10,184 mOsm/360.48 mOsm/L = 28.25 L

So with a hypertonic solution, we get an expansion of the EC by 14 L (100% increase in volume = 2X) and a contraction of the IC by 6 L (25% decrease in volume= . 25 or 1/4). Both compartments increase their osmolarity from 280 to 360 (a difference of 80 = 29% increase creating a hyperosmolar condition= increase in concentration of solute). This is a state of hyper-osmotic dehydration of which the body needs to balance the solute load to come back to equilibrium. We will see that is can be managed by sweating and kidney function with the result of decreasing the temperature of the body during exercise preventing heat exhaustion or heat stroke.

Chapter 12

The Water Overload State (Hypotonicity)

Osmolarity is determined by the amount of solute (Na+ concentration) divided by the volume of E.C. fluid.. Water playing a big role in this process. The water is determined by the thirst and renal excretion of water. We will review the feedback mechanism involving the brain and kidney, also the thirst and salt appetite mechanism.

What happens when the body has too much water in the system? What is the manner in which the body controls excess water on a normal basis? The answer is the feedback mechanism between the posterior pituitary gland and the distal tubules and collecting ducts of the nephron. In the brain there is an area called the Hypothalamus. Inside the Hypothalamus is a nuclei called the supraoptic nuclei. This area is home to cells called osmoreceptors. These cells sense the osmolarity of the E.C. fluid by way of the Na+ concentration. In our case, the Na+ concentration is low, osmolarity is low, therefore the receptor down regulates it's activity. ADH (vasopressin) is not released by the gland and the kidney is not permeable to water for reabsorption,

Osmoreceptor Feedback Mechanism

ECF : Low [Na+]
        Low osmolarity
Hypothalamus : Supraoptic nuclei
                Osmoreceptor
                Neuron
Posterior Pituitary Gland : ADH
                Epithelial cell
Kidney : Epithelial cell
                Distal tubule & Collecting Duct
                Water permeability : Reabsorption

With this mechanism, the kidney creates a range of dilution from 50 mOsm/L (dilute =1/6, 20 L/day) to 1400 mOsm/L (concentrate=5X) [63].

So in the presence of ADH, water is reabsorbed in the distal tubule and collecting duct of the kidney. This can happen when a hypertonic solution is ingested such as Cranapple. But if a hyptonic solution such as water is ingested, ADH is absent permitting the kidney to excrete dilute urine. Let's look at the effect of drinking hypotonic solution (the absence of ADH).

Renal Response to Water [64]

	mOsm/L	Urine (ml/min)	Urine (mOsm/min)
Time 0 (min)	300 (Blood) 600 (Urine,6X)	1	0.6
45 mins.	280 (Blood)	6	.61
	100 (Urine)		

At 45 minutes:

*12A Drink 1 Liter of H2O

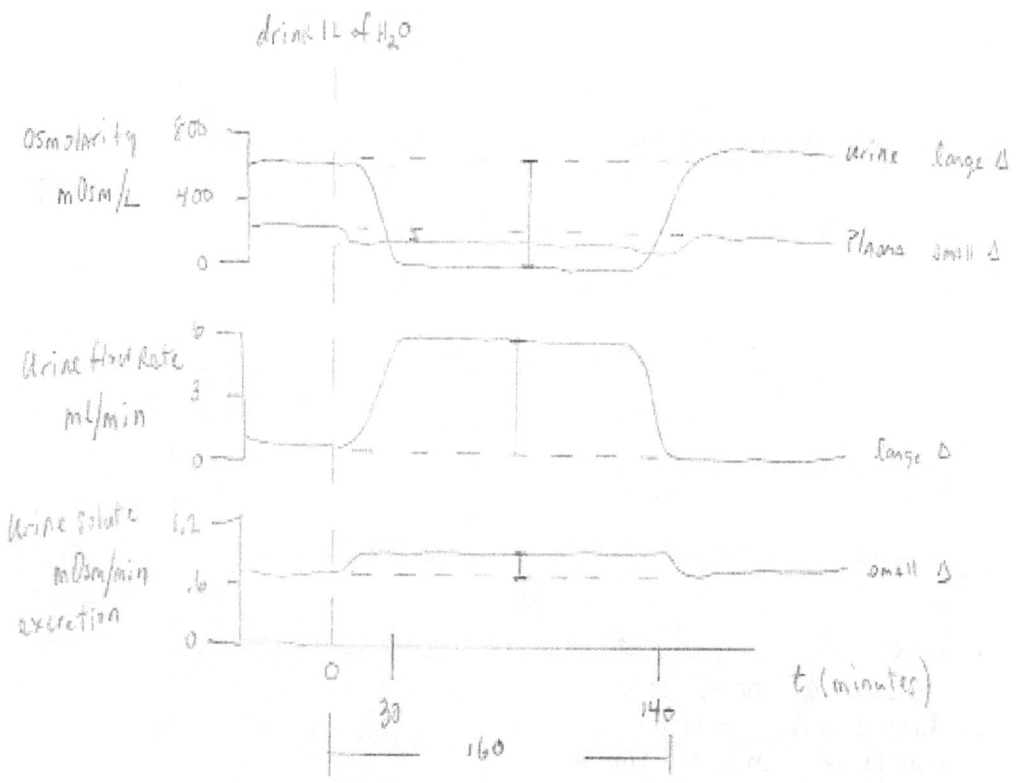

Note the very little change in solute excretion by the kidney and plasma osmolarity. The conservation of Na+ is vital to cell function. The 45 minute mark is important as the start point for making massive amounts of urine volume, but notice the duration of the diuresis, 150-45=105 minutes, approximately a little under 2 hours (1hr + 45 mins). This represents a good point to consider emptying the bladder before a sporting event. So oral hydration (pre event) should start 2.5 hours prior to the event with voiding 30 minutes prior to the event.

At the level of the kidney, The normal structure of the kidney is Nephron (Functional Unit of the kidney) [65]:

*12B The Glomerulus

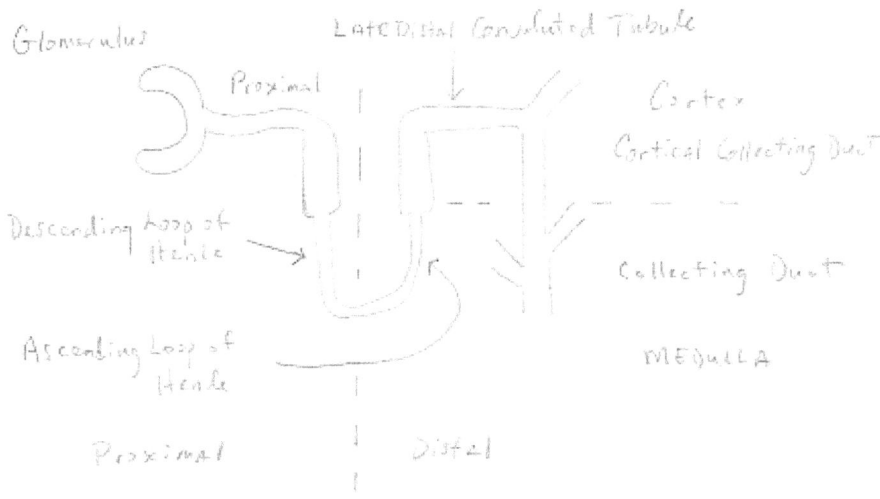

From proximal to distal:

1. Proximal Tubule
2. Descending Loop of Henle
3. Ascending Loop of Henle
4. Late Distal Convoluted Tubule
5. Cortical Collecting Duct
6. Collecting Duct

The kidney is grossly divided into cortex (outer) and medulla (inner) [66]

*12C Gross Kidney

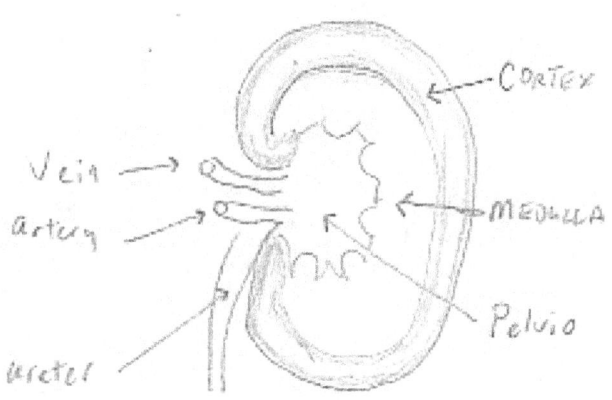

The Nephron [67]:

Functional Unit of the Kidney

Vasa Recta "U" shape

*12D Vaso Recta

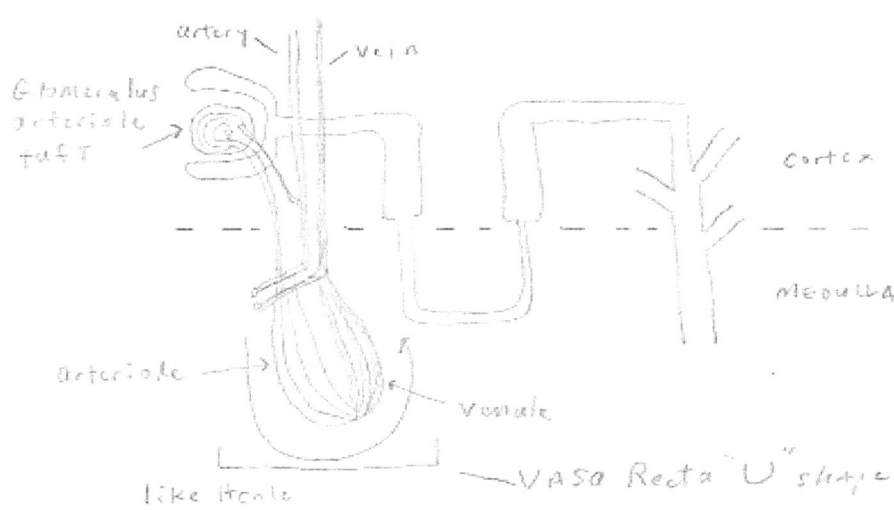

Histologic section of Medulla (Micro) [*68*]:

*12E Medulla

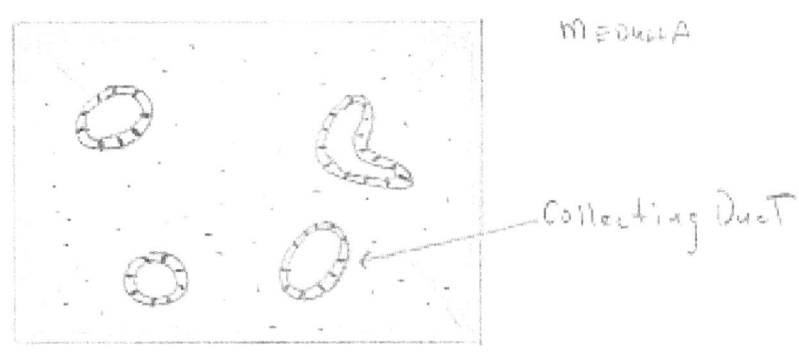

Histologic section of the cortex (Micro) [*69*]:

*12F Cortex

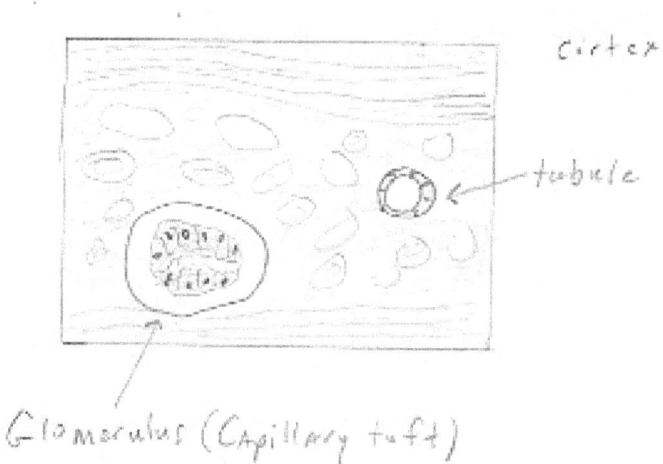

The changes occurring after drinking water (hypotonic solution).

Renal Response to Water Load (Na+ & H2O Activity)

Kidney	Cortex (300 mOsm/L)	Medulla (600 mOsm/L)
Proximal Tubule (isosmotic)	Na+ (reabsorbed **Actively**) H2O (reabsorbed **Osmosis)**	
Descending Loop		H2O(reabsorbed Osmosis)
Ascending Loop **Thick** segment	Impermeable to H2O	**Na+** (reabsorbed Actively) Na+ (reabsorbed Actively)
Late **Distal** (*ADH*)		Na+ (reabsorbed Actively)
Collecting (*ADH*)		Na+ (reabsorbed Actively)

Change in osmolarity: 300 –> 50 mOsm/L
Net result: Dilute urine
Water excess corrected.

*12G Renal Response to Dilute H2O

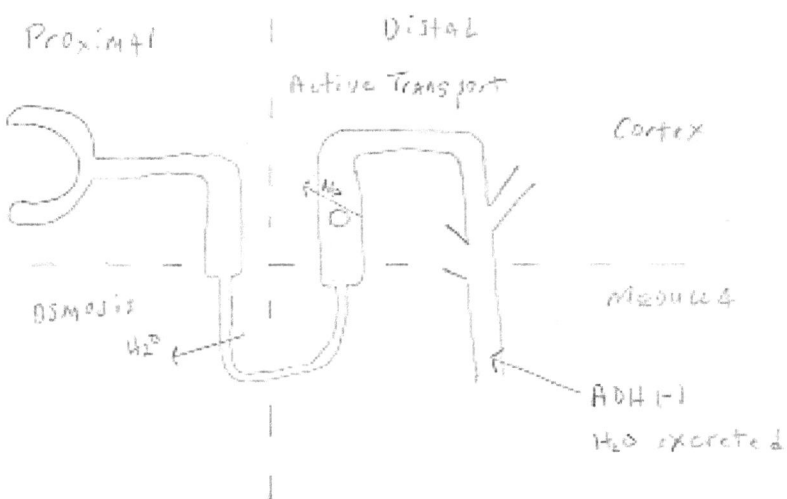

Renal response to Water (mOsm/L)

Kidney	Cortex (300) Isosmotic	Medulla (600) Hyper-isomotic
Proximal tubule Isosmotic (300)	300	
Descending Henle Hper-isomotic (600) (2X)		4-600
Early/Late Distal (100) (1/3X) Hypo-isomotic ([-] ADH)	300	
Cortical Collecting (70) Hypo-isosmotic ([-] ADH)		400
Collecting (50) (1/6X) Hypo-isosmotic ([-] ADH)		600

All the mOsm/L are compared to plasma (300).

The Ascending limb is impermeable to H2O. From the Proximal to the Ascending Limb, ADH is not involved in the process. This is what normally happens. ADH makes the Distal and the Collecting ducts permeable to water (reabsorption of water). In this case of drinking hypotonic fluid, ADH is absent (negative feedback).

Physiologic response to water

ECF (Na+)	Osmoreceptor	ADH	Distal/Collecting
Low	Off	Off	Off

Result: Hypo-isomotic urine output (more water that solute=Na+).

Chapter 13

The Dehydrated State (Hypertonicity)

Now what happens when an athlete drinks hypertonic solution such as Cranapple. The Cranapple has an osmolarity of approximately 800 mOsm/L. This means that the kidney will need to concentrate the urine; the kidney will excrete more solute than water. There is 2 facts in which we need to know before doing the calculation. The maximum concentrating ability of the kidney is between 1200-1400 mOsm/L (4-5X greater than plasma (300)) [70]. Also the obligatory solute excretion of 600 mOsm/L of solute/day. This equals .5L/day, the obligatory urine volume, of water needed to excrete metabolic waste. This is calculated using [600 mOsm/L] / [1200 mOsm/L] = .5L/day. We are now ready to calculate the changes that will occur by drinking 8 Liters (approximately 2 Gallons) of Cranapple. [800 mOsm/L] / [600 mOsm/L] = 1.33 X which means that it will result in a net fluid loss of 0.33 L. 0.33 L * 8 = 10.67 L total. This is the amount of water needed to drink due to the obligatory lost created by drinking Cranapple. This is a fairly large amount of water. When we divide this over 3 days prior to game time; this brings it down to 10.67 L/3 days = 3.56 L/day; much more reasonable.

Generalized Formula

Osmolarity of Drink/Obligatory Osmolarity of Kidney = Magnitude of Comparison

Magnitude -1 = Fractional Difference

Fractional Difference * Total Volume = Total Water Replacement

The net result of drinking hypertonic solution is to dehydrate the body. We do this in sports to load the body of Na+ because of the expected losses during the game, which is very important for neuromuscular function. To compensate for dehydration, we replace the water deficit with water. Let's look at this in more detail.

When we load the body with solute, the kidney needs to concentrate the urine to compensate for the change. 2 mechanisms come to light: 1. ADH and 2. Renal Medulla hyperosmolarity. The Medulla mechanism is called the countercurrent mechanism. It is made possible by both the specialized nephrons, juxtamedullary, and vasa recta, peritubular capillaries. The force behind it all is osmosis.

Countercurrent Mechanism[71]

Normal cell interstitium osmolarity=300mOsm/L
Renal Medulla interstitium osmolarity=1400
   Active transport by Na+ pumps in thick ascending limb of Loop of Henle and collecting ducts.
   Facilitated diffusion of urea from collecting ducts
   diffusion of water (very small)
Functional character of the Loop of Henle:
   Thick ascending=Na+ AT out of tubule=200 mOsm gradient
   Thin ascending=Na+ diffusion (passive) out of tubule
   Descending limb=H2O osmosis out of tubule
The steps of mechanism:
   1. Proximal–>Distal tubule osmolarity=300
   2. Thick segment turned on creating 200 mOsm gradient. Paracellular diffusion limited.
   3. Descending limb H2O out by osmosis.
   4. Thick segment turned on again.
   5. Descending limb H2O out by osmosis.

   Paracellular means the flow of Na+ between the cells.

   The Proximal tubule primes the Distal segment; supplies the Na+ load. With the repetitive action of the Thick ascending limb AT of Na+ out of the tubule with osmosis of H2O out of the Descending limb with the creation of a 200 mOsm gradient, a countercurrent multiplier mechanism is established. The Thick ascending limb creates a 100 mOsm gradient, but on the Descending limb side, 100 mOsm osmotic gradient is established. Adding these together is the 200 mOsm gradient. It is called countercurrent due to the fact that Na+ flow opposes H2O flow. It multiplies on itself.

*13A Countercurrent Mechanism by the #s

Countercurrent Mechanism by the #s [72]

	Proximal	Distal	Medulla
Step 1	300	300	300
Step 2	300	200	400
Step 3	400	200	400
Step 4	400	400	400
Step 5	400	300	500
Step 6	500	300	500
Step 7	500	500	500
repeat 4,5,6,7-->1200			

H2O   Na+

100   ➤   ◄——   100

200

To complete the picture, we need to add the role of ADH. The last 3 parts of the nephron, distal, cortical collecting and medullary collecting ducts play vital role in a active transport and ADH. The urine is diluted when it reaches the distal tubule because of the active transport of Na+ out of the tubule. In the cortical collecting tubule, in the presence of ADH, H2O is reabsorbed and quickly swept sway by the peritubular capillaries. This occurs in the cortex of the kidney keeping the medulla hyperosmolar. Same occurs in the medullary collecting tubule in the presence of ADH, but this time very little H2O remains for absorption. Net result is a concentrated urine, more Na+ than H2O. So when an athlete ingests a hyperosmolar drink such as cranapple, he loads this body with the nutrient, such as Na+. This is dehydration making his ECF hyperosmolar. The kidney responds by excreting a concentrated urine, urine Na+ lost. This modified by the sweating process during competition and the drinking of pure water. To concentrate the urine requires 1. energy in the form of ATP for the Active Transport of Na+ and 2. the secretion of ADH.

Renal response to Na+ (mOsm/L)

Kidney	Cortex (300) Isosmotic	Medulla (600) Hyper-isosmotic
Proximal tubule Isosmotic (300)	300	
Descending Henle Hper-isosmotic (600) (2X)		4-600
Early/Late Distal (100) (1/3X) Hypo-isosmotic ([+]ADH)	300	
Cortical Collecting (300) Hypo-isosmotic ([+] ADH)		400
Collecting (1200) (4X) Hypo-isosmotic ([+] ADH)		1200

Physiologic response to Na+

ECF (Na+)	Osmoreceptor	ADH	Distal/Collecting
High	On	On	On

Result: Hyper-isosmotic urine output (more Na+ than H2O)

In the presence of hyperosmolarity, ADH does the following:
1. Hyperosmolarity means Na+>H2O=water deficit.
2. ECF mOsm increased.
3. Osmoreceptors activated via shrinkage. Cells located in the anterior hypothalamus near supraoptic nuclei. Shrinkage cause nerve signal generation.
4. ADH increased in the Posterior Pituitary Gland. ADH stored in granules in the nerve endings.
5. ADH increased in plasma.
6. H2O increased reabsorption by the kidney by the Distal and Collecting ducts.
7. H2O decrease secreted by the kidney (Concentrated=Na+ > H2O).

8. H2O deficit eliminated (Negative feedback).

Under maximum concentrating ability, urea forms 50% (600 mOsm/L) of the hyperosmolar medullary intersitium [73]. Urea is reabsorbed passively and by facilitated diffusion. ADH has a profound influence on urea absorption in the medullary collecting duct. This has been described as the re-circulation cycle of urea. Passive diffusion of urea occurs in the proximal, thin descending and thin ascending segments of the Loop of Henle. At the medullary collecting tubule, urea udergoes facilitated transport by urea transporters, UT-AI, which is activated by ADH. Therefore there is a cycle of urea through the medulla intersititium from collecting tubule to thin segment, adding to the hyperosmolarity of the medulla.

The vasa recta also preserves the hyperosmolarity of the medulla by having very low blood flow and by acting as countercurrent exchangers. The exchanger mechanism is largely due to the "U" shape of the vasa recta.

Chapter 14

ADH & The Thirst Center (Conscious Control of Dehydration)

Water balance is controlled by a physiologic control system located in the hypothalamus [74]. Body water homeostasis is achieved by balancing intake and output (losses). The thirst center is located in the anteroventral hypothalamus. It's major stimulus is osmolality. The **osmoreceptors** do the sensing. Also hypovolemia (low blood volume) has an effect mediated by arterial baroreceptors and the renin-angiotensin system. ADH is made from the neurosecretory cells located in the supraoptic and paraventricular nuclei in the hypothalamus [76]. Then is is transported in granules down axons to the posterior pituitary [75]. ADH is stimulated by a variety of stimuli: osmolality, hypovolemia, renin-angiotensin system, hypoxia (low oxygen), hypercapnia (high CO2), hyperthermia (high temperature), pain and drugs [77]. Blood volume and Blood pressure control have priority over osmolality.

*14A Pituitary Gland

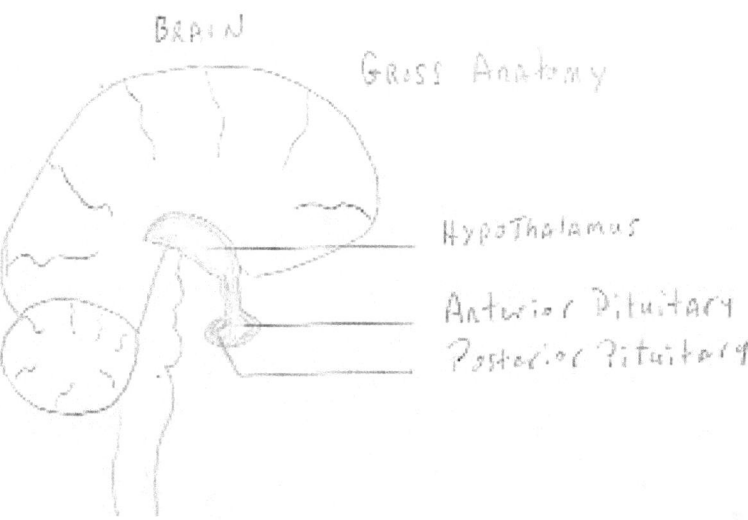

*14B Axis

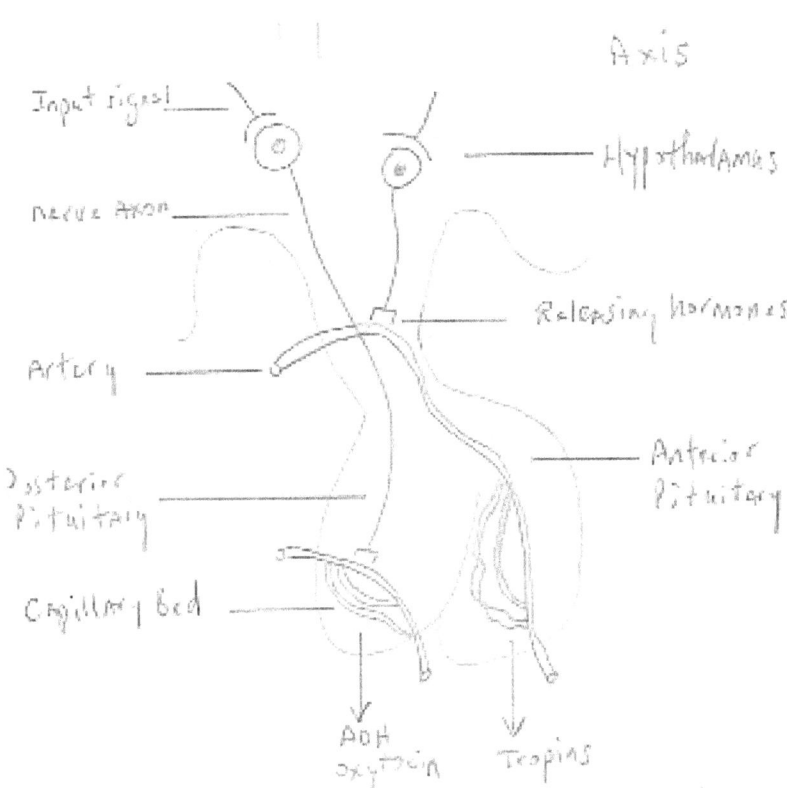

*14C Nuclei

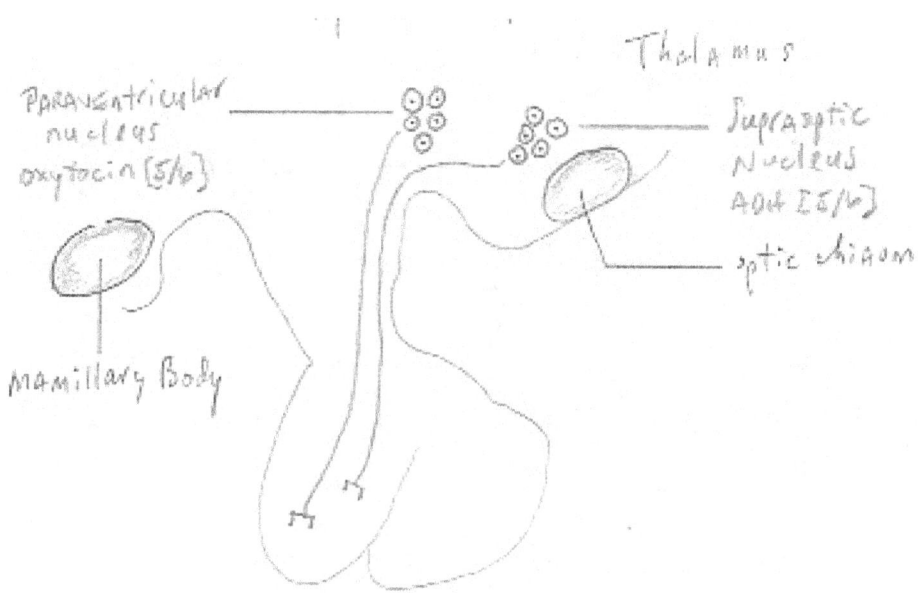

5/6 is the fraction of the primary hormone produced: 5/6 ADH + 1/6 Oxytocin [*78*].

Both nuclei produce ADH.

*14D ADH Formula

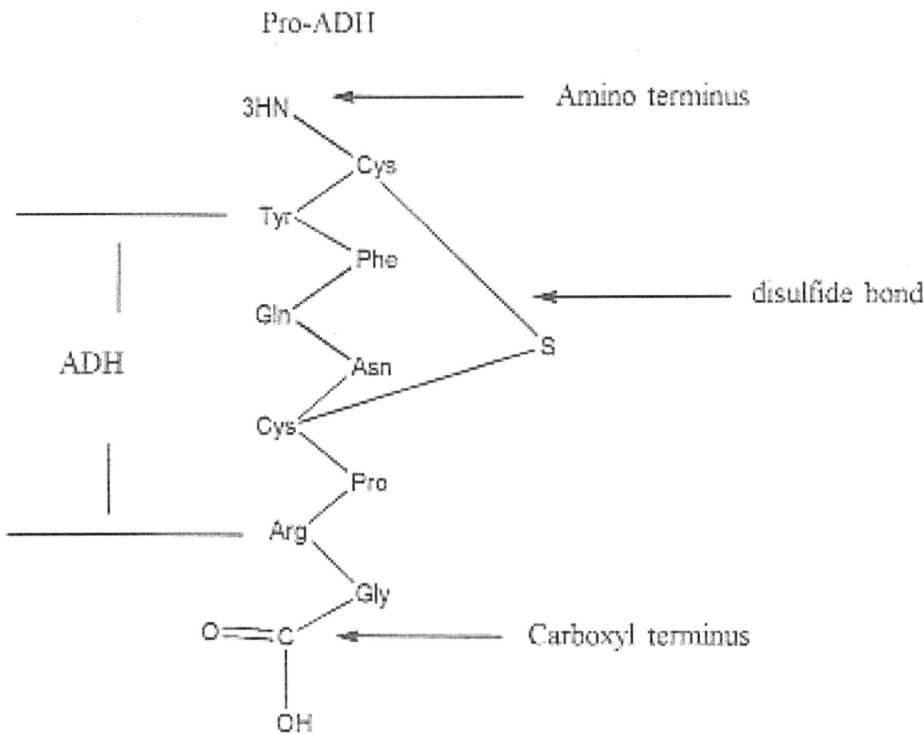

*14E ADH and Thirst Center

Gross Surface Anatomy

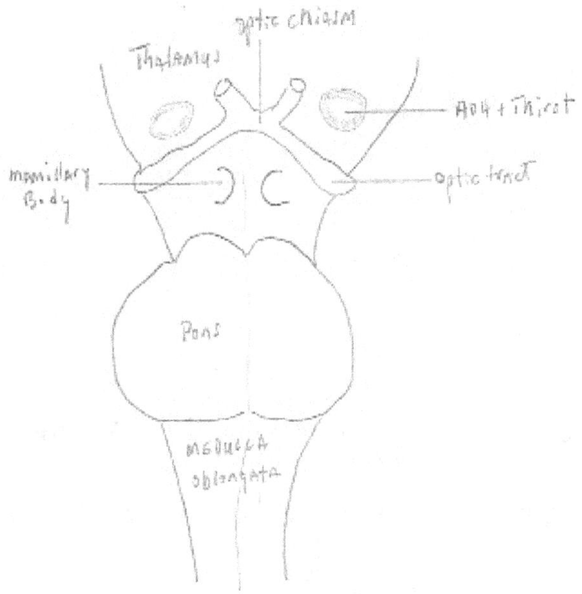

optic chiasm

Thalamus

ADH + Thirst

mamillary Body

optic tract

Pons

medulla oblongata

ADH and Thirst are closely related both in anatomical location and physiologic control.

(Ventral View) [79].

*14F Supraoptic Nucleus

Cut Section —— III Ventricle

—— Paraventricular nucleus

—— Supraoptic nucleus

—— optic chiasm

Thirst Center: Anteroventral wall of the **3rd Ventricle** + Anterolaterally in the **Preoptic Nucleus**.

Cross Section   ADH and Thirst Centers on both sides of 3rd Ventricle (AV3V) [80].

Known Stimuli and Inhibitors of Thirst and the Release of ADH [81]:

Stimuli: Osmotic: Loss of ICF
              Nonosmotic: Decreased BP
                           Decrease Artery Wall Tension
                           Pain
                           Psychosis
                           Fever
                           Drugs
                           Hypoxia or Hypercapnia
                           Stimulation of the Renin-Angiotensin System

Inhibitors: Osmotic: Increase ICF
              Nonosmotic: Increase BP
                          Increase Artery Wall Tension
                          Emotional Stress
                          Low Temperature
                          Drugs
                          Inhibition of the Renin-Angiotensin System

Stimuli and Inhibitors are opposites in action.

BP and Blood Volume takes priority in the control mechanism. Maintaining the pressure and volume of the vascular compartment of the E.C.F. takes priority over osmolarity.

The median **Preoptic** nucleus is connected to the **Supraoptic** nuclei and **blood pressure control centers** in the medulla of the brain. These areas lack the blood brain barrier located
elsewhere in the brain. Making the ADH response rapid.

The arterial baroreceptor **reflexes** and cardiopulmonary reflexes play a central role in control of the homeostasis of the E.C.F.. They originate in the aortic arch, carotid sinus and cardiac atria. The nerve impulses are carried by the 9th and 10th cranial nerves, glossopharyngeal and vagus. They connect to the nuclei of the **Tractus Solitarius**. From here, they end at the **hypothalamus**.

Missing stimuli on this list includes:
Dry mouth and esophagus
GI and pharyngeal stimuli

*14G ADH Negative Feedback

Negative Feedback Mechanism

OSmoreceptors   Thalamus

+ ∆ osmolarity   ↓

Nuclei   Synthesis ADH

↓

Posterior Pituitary
Stores ADH   ADH

+/- ∆ Volume

Stretch Receptors   Renal Collecting Duct
Atria   H₂0 reabsorbed
Vena Cava
Carotid Sinus

*Volume vs. osmolarity [82]

To Hypothalamus–>Posterior Pituitary Gland–>ADH–>Renal Collecting Duct

H20 reabsorbed by osmosis (passive event).

ADH with the renal medulla (counter current mechanism) compensates for the hypertonic state (concentrated urine).

ADH –> cell membrane of kidney epithelial cell –> cAMP –> phosphorylation (add PO -4) to proteins –> microtubule behavior [83].

Microtubules possibly assist in the development of the topography of the surface epithelium (microvilli + pores=water channels).

*14H Microvilli and H2O Pore

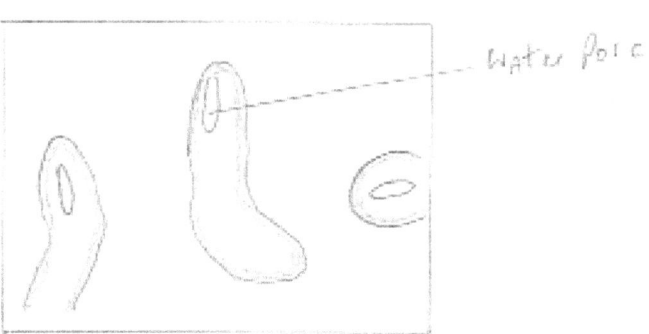

ADH increases the # of pores [84].

*14I Osmoreceptors

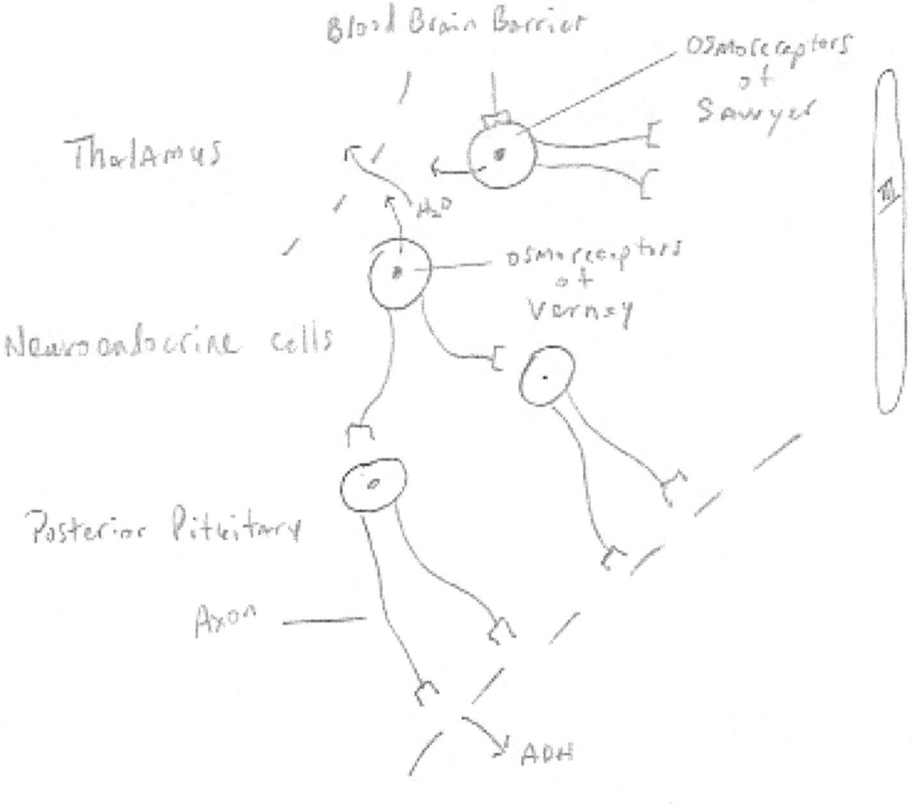

Osmoreceptor of **Verney**: senses osmolality.

Verney –> Magnocellular neuron –> ADH [85].

Osmorecepton of Sawyer: senses noxious and osmolality.

Thirst –>Behavior–>Magnocellular neuron–> ADH

Noxious–>Magnocellular neuron–>ADH

Blood Brain Barrier in this area is much more permeable than the other membranes of the body.

Chapter 15

Sweating (The Cooling Process)

Strenuous exercise can cause a temporary rise in body temperature to a range of 101 to 104 degrees F [86]. The heat is created in the deep organs of the body and transferred to the skin surface. The transfer is mainly due to blood flow. This flow can vary from 0 to 30% of the total cardiac output [87].

*15A Ways of Losing Heat

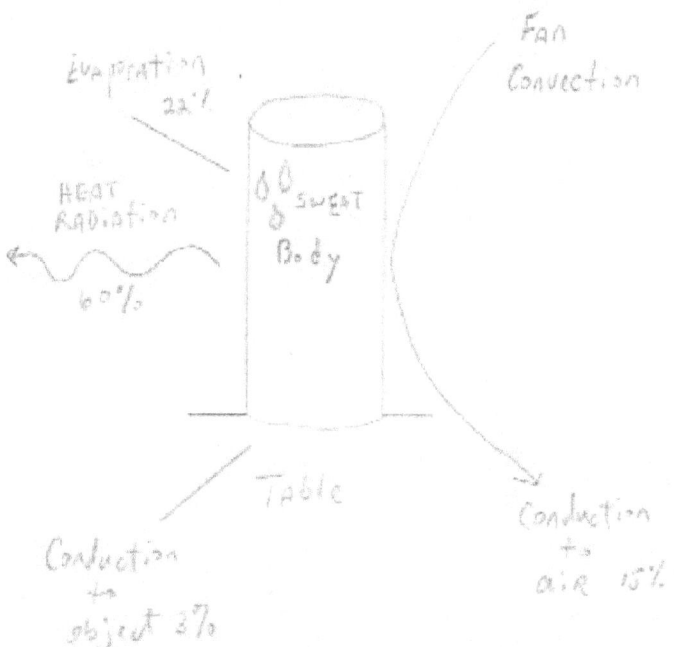

Fan can increase the heat lost [88].

Well ventilated clothing can increase heat lost.

Being wet can increase heat lost.

.58 kCal of heat is lost/1 gram of H20 [*89*].

Acclimatization takes 4-6 weeks.

Unacclimatized: NaCL=50-60mEq/L lost
Normal: 1L/hr sweat produced.
Exercise: 2-3 L/hr. sweat produced.
15-30 gms salt/day (1st few days)
K+: 1.2X plasma lost

Acclimatized: NaCL=5mEq/L lost.
Salt lost: 3-5 gms./day
Aldosterone effect

*15B Sweat Gland

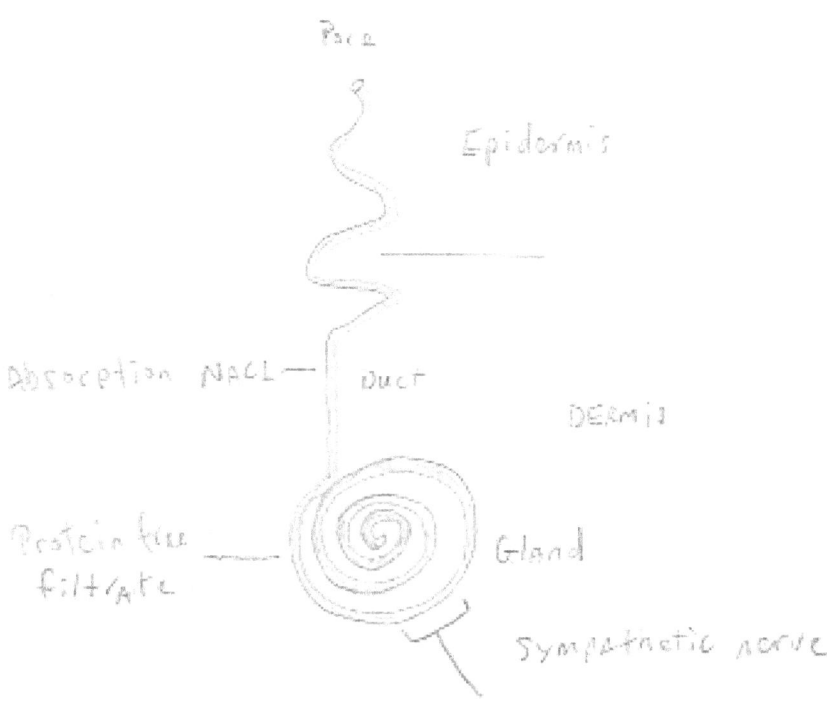

At the duct, NaCl is ½ of plasma concentration [*90*].

At the gland, Na+ = 142 mEq/L, Cl- = 104 mEq/L (=plasma).

A High School football game takes about 1 hour of play time. The entire event is approximately 4 hours.

Using the 4 hour mark as a maximum:

> . an unacclimatized player:
> he will produce: 4 hrs.*3 L/hr.=12 L of sweat.

> This will contain: 12L*60 mEq/L of NaCl= 720 mEq.
>   1.2X plasma*5 mEq/L=6 mEq/L * 12L=72 mEq.

K+

> an acclimatized player: these loses are approximately 1/10 in amount.

4 hour Football Game

Lost	NaCl	K
Unacclimatized:	720 (mEq)	72 (mEq)
Acclimatized:	72	7.2
Cranapple minimum replacement		
Unacclimatized:	2.160 L	1.152L
Acclimatized:	.2160L	.1152L

720 mEq lost/80 mEq/240 ml=2.160L
72   "   /15 " / "   =1.152L
1 mEq= 1mg

So when we pregame hydrate, we Na+ load which creates a dehydrated state. The kidney concentrates the urine by excreting the Na+ load. Sweating initially excretes the Na+ load also, but over time, aldosterone decreases this effect. The sweating becomes the radiator for the body by dissipating heat preventing organ damage and death.
K+ depletion of the body is unusual, but can occur in total body dehydration (I.C.F.). The unacclimatized person is most susceptible.

1ml=1g   12L=12000 gms.* .58 kCal = 6960 Cal of Heat./4 hrs.=1740 Cal./hr

Facts on Heat Stroke:

Heat exhaustion –> Stroke (medical emergency)
104° F is high-106°F can be deadly.

Symptoms:
Headache
Fainting or dizziness
Red skin color –> pale, blue color
Fever
Seizure
Nausea/Vomiting
Blindness
Unconsciousness

Signs:
Early: Tachycardia/Tachypnea –> Hypotension

First Aid:

911 –> Hospital
Indoors/shade
Remove clothing
Bath **cool** water
Hyperthermia vest
Cold compresses to torso, head, neck and groin
Fan/air conditioning unit
Hydration = drink/IV
CPR (cardiac arrest)
Recovery position (air way maintained)

Prevention:
Light, loose-fitting clothing
Wide-brimmed hats with vents
Hot weather alert
Color of urine: Dark=dehydration
Clear=like spring water (hydrated)

OSHA Heat Stroke Quick Card for prevention:
Know Signs/Symptoms
Block heat sources

Use cooling fans/air-conditioning
  Rest regularly
  Drink water (1 cup/15min.)
  Wear lightweight, light colored, loose-fitting clothes
  Avoid alcohol, caffeinated drinks or heavy meals.

Chapter 16

Acclimatization: The Aldosterone Effect

Prior to acclimatization, SWR are very high producing tremendous Na+ and H2O loses. At acclimatization, these rates decrease almost 6 fold. This is due to the Aldosterone effect. Aldosterone conserves Na+ at the expense of K+. Fluid and electrolyte replacement must reflect this change to prevent deficits in Na+, K+ and H2O.

Environment: Hot and Humid

Data Table [91]

	Normal, unacclimatized	1st 6 weeks	Acclimatized
SWR	1L/h	2-3 L/h (*G)	
Heat dissipation		>10X BMR	
Na+ loss		15-30 g/d	3-5 g/d (A)
Weight loss		5-10 Lbs./h	

* (G) Sweat gland adaptation
  (A) Aldosterone effect
3% weight loss effects performance
5-10% weight loss cause symptoms, dizziness, cramps, and nausea.

These numbers represent ranges. Measurements can be made to be more precise and individual.

Aldosterone is made in the outer most layer of the adrenal gland [92].
The gland sits on tope of each kidney.

*16A Adrenal Gland

Cortex 3 Zonas

Glomerulosa (Aldosterone)

fasciculata (Cortisol)

Reticularis (Androgen)

Medulla (Catecholamines)

*16B Aldosterone Formula

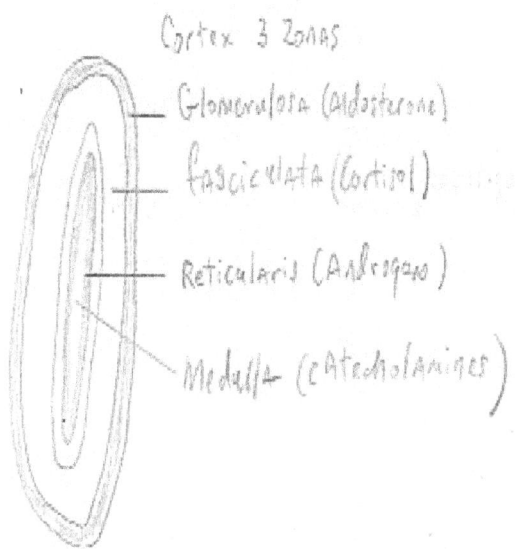

cholesterol

Progesterone

Aldosterone

*16C Steroid Activity

Glucocorticoid + Mineralocorticoid Activities

Steroid	?tAms [] ua/?one	Amt Secreted mg/d	Gluco	Mineralo
Adrenal: Cortisol	12	15	1	1
Aldosterone *	.006	.15	.3	3000
Deoxycorticosterone	.02b	.2	.2	100
Synthetic: Cortisone			1	1
Dexamethasone			30	
9α-fluorocortisol			10	125

* Aldosterone is a very strong mineralocorticoid. (3000) [93].

Actions of Aldosterone:

> Renal: Principal cells (collecting tubules*/distal/collecting duct)
>   Absorb Na+/Excrete K+
> Sweat Gland/Salivary Gland/Large Intestine
>   Absorb Na+/Excrete K+

Cellular Mechanism:

> Lipid soluble, passage through cell membrane
> receptor protein in cytosol
> Into nucleus to induce DNA–>RNA

make 1. Enzymes (Na+-K+ adenosine triphophatase) (2. Transport proteins (epithelial Na+ channel proteins)
> Start time is 45 minutes, maximum is 3 hours [94].

Regulation of Aldosterone:

> 1. Increase K+ in E.C.F.
> 2. Increase angiotensin II of renin-angiotensin system causes slight decrease in Aldosterone.
> 3. A.C.T.H. needed but not in rate control.

4. Increased Na+ in E.C.F.

Chapter 17

Field Applications 1,2

I

Environment: Hot and Humid

Source of Player: National **Collegiate** Athletic Association Division II Team

Football Player Characteristics: Larger
                Not as aerobically fit
                Higher % body fat : Retains Heat
                Higher BSA : More sweat glands
                Insulating Equipment: Less lose of Heat
                    Prohibit sweating
                Higher Sweat Rates:　> 2L/hr.
                Sweat lost maximum during preseason
                    Not acclimated
                Average Sweat Losses > 9.4 L/day
                    8L during sessions
                    Consumed large volumes of fluids
                    during exercise but still were
                    dehydrated.
                Fluid replacement required in between
                    sessions.
                    3L between Am-PM
                    3L after PM
                Daily fluid requirement = 12 L/day
                Na + supplementation critical to prevent
                    chronic dehydration.

                Na+ supplementation days prior to exercise
                    is helpful
                Posture (sitting on the bench) causes plasma
                    volume contraction

                Plasma Volume declined 5% during exercise

    Fluid Replacement guidelines: Calculated:

130% * sweat lost volume - Fluids consumed

Fluid replacement after sessions [95]:

*17A Fluid Requirements

Fluid Requirements

	Between Am-Pm	Night Before	24 hr
Football (FB)	3.1 (L)	3.0	12.2* (3x)
Cross Country (CC)	1.7	1.7	4.6

Following fluid replacement guidelines with these given losses, could promote hyponatremia (low Na+). Electrolyte replacement is important.

*17B Plasma Volume Change

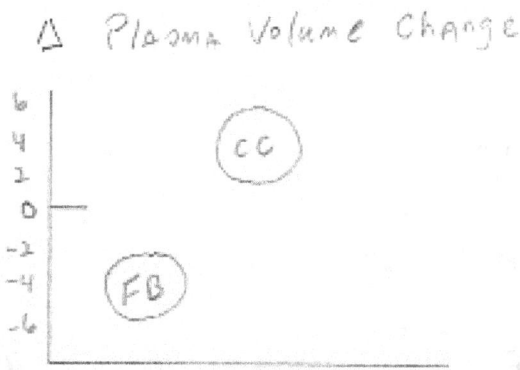

FB: Plasma volume lost (-) is indicative that football players are dehydrated [96].
During Sitting/Bench time, dehydration is unexpected, but occurs just as if the player was in the game.

*17C Urine Specific Gravity

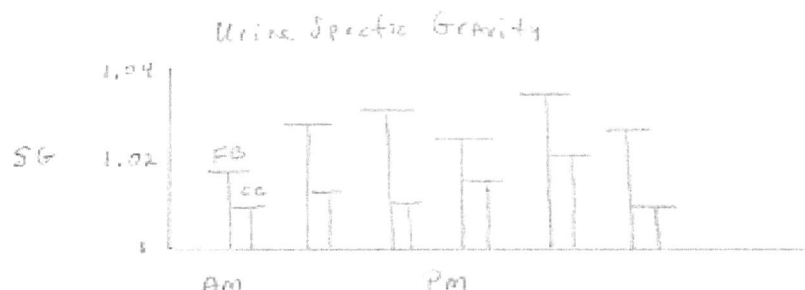

Urine is concentrating, minimizing H2O lost in order to maintain plasma volume.Na+ lost is mainly from sweat.
Kidney will reabsorb Na+ and H2O and concentrate the urine with other solutes [*97*].

Factors predisposing athletes to heat intolerance:
  History of heat related problems
  Drug & alcohol abuse
  High BMI
  Obesity
  Poor aerobic fitness
  Dehydration
  Lack of acclimatization

Factors varying sweat rates:
  Ambient temperature
  humidity
  Air movement
  Exercise intensity
  Insulating clothing and equipment
  Body size

Mild Na+ depletion:

      Nausea
      Fatigue
      Headache
      Muscle twitching

Severe Na+ depletion:
      Seizures: Requiring IV Hypertonic solutions slowly.

Factors affecting plasma volume:
      Physical fitness
      Hydration status
      acclimatization
      posture
      type of exercise

Summary: H2O and Na+ hydration is very important during acclimation period (4-6 weeks). Proper attire for the given environment. Early recognition and re-mediation of symptoms when they present. Larger volumes of input and output are involved with football players which has the potential tendency towards hyponatremia. Football players usually have their plasma volume contracted potentially leading to pre-shock and eventually death. Larger of volumes of fluid intake can be tolerated by the players given the intermittent nature of their activity.

II

Environment: Hot and Humid

Source of Player: **NFL** Lineman and Backs

                                     Linemen: Larger Mass [98]
                                     Larger BSA
                                     Smaller BSA/Mass

*17D Physical Characteristics

Physical Characteristics Lm And B

Physical	Linemen	Backs
Age (yr)	27	25
Ht (cm)	193	184
MASS (kgs)	133	89
BSA (m²)	2.6	2.1
BSA/m (cm²/kg)	195	236

Calculations:

Sweat rate =

 [(prepractice mass (kg) - postpractice mass (kg))] - postpractice urine volume (L) + Fluids consumed during practice (L) /

length of practice session (h).

BSA (m^2) = .020247 * Height (m)^ 0.725 * Weight (kg) ^ 0.425

*17E Sweat Rate and Fluid Rate

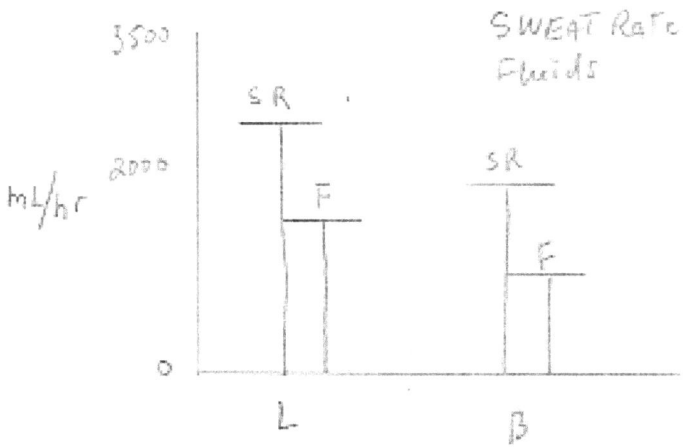

Linemen have higher sweat rates and less fluid consumption than backs [99].

*17F Sweat Lost and Fluids

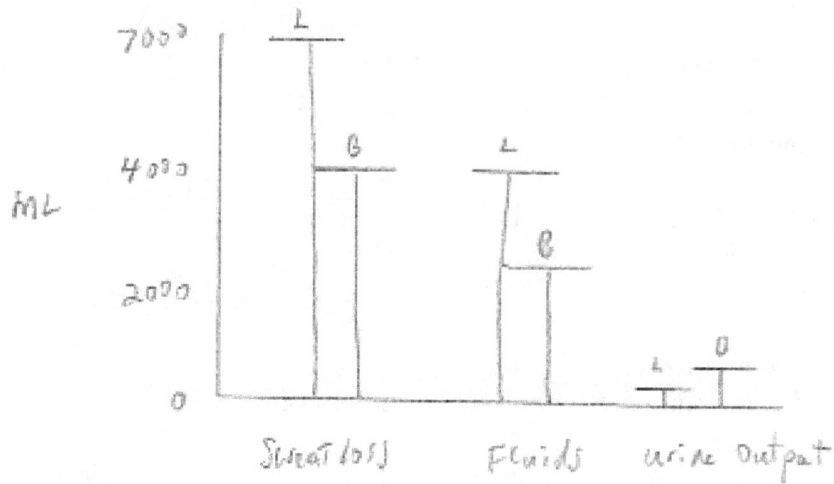

Linemen more dehydrated because of lower urine output [100].

Linemen activity behavior: Run very short distances
                                       Lower to the ground
                                       Lower stance
                                       Close to hot bodies
Lower evaporative heat lost
No ventilation (No fan effect/wind)

Na+ Lost can be as high as 11g/d. [101]

For a given volume lost, there is less Na+ lost. The Na+ concentration is maintained. Replacement of volume with hypotonic solutions could lead to hyponatremia.

Summary: Linemen are more prone to dehydration given their body characteristics, but also to their playing behavior and job circumstances.

Endnotes

1. Anderson, Douglas M., Keith, Jeff, Novak, Patricia D., Elliot, Michelle A., **Mosby's Medical, Nursing, & Allied Health Dictionary**, 6th Edition, (St. Louis, Missouri, 2002), 824.
[Definition: Homeostasis]

2. Guralnik, David B., **Webster's New World Dictionary of the American Language**, 2nd Edition, (New York, Simon & Schuster Inc., 1984), 42.
[Definition: Automaton]

3. **The Harriet Land Handbook**, A Manual for Pediatric House Officers, 9th Edition, (Chicago, Year Book Medical Publishers, Inc., 1981), 304.
[Normal Range of Na+ in blood]

4.. **http://www.bio-medicine.org**/Biology-Definition/Life/#A_conventional_definition[Definition: Life]

5. Heller, Craig, Orians, Gordon, Purves, William K, Sadava, David, **Life, the Science of Biology**, 6th Edition, (Sinauer Associates, Inc., 2001), 694.
[Diagram: Homeostasis in functional systems]

6. Wikipedia, 2009.
[Table: Normal Values in Humans]

7. Heller, Craig, Orians, Gordon, Purves, William K, Sadava, David, **Life, the Science of Biology**, 6th Edition, (Sinauer Associates, Inc., 2001), 21.
[Diagram: Electrons and Shells]

8. Heller, Craig, Orians, Gordon, Purves, William K, Sadava, David, **Life, the Science of Biology**, 6th Edition, (Sinauer Associates, Inc., 2001), 21.
[Table: Bonds and Interactions]

9. Heller, Craig, Orians, Gordon, Purves, William K, Sadava, David, **Life, the Science of Biology**, 6th Edition, (Sinauer Associates, Inc., 2001), 23.
[Diagram: Polarity]

10. Heller, Craig, Orians, Gordon, Purves, William K, Sadava, David, **Life, the Science of Biology**, 6th Edition, (Sinauer Associates, Inc., 2001), 23.
[Diagram: H-bond]

11. Heller, Craig, Orians, Gordon, Purves, William K, Sadava, David, **Life, the Science ofBiology**, 6th Edition, (Sinauer Associates, Inc., 2001), 26.
[Diagram: Tetrahedron Geometry]

12. Kotz, John C., Treichel, Paul Jr., **Chemistry & Chemical Reactivity**, 3rd Edition, (Saunders College Publishing, 1996), 60.
[Table: H2O Properties]

13. Heller, Craig, Orians, Gordon, Purves, William K, Sadava, David, **Life, the Science of Biology**, 6th Edition, (Sinauer Associates, Inc., 2001), 21.
[Diagram: Na+ : Electrons and Shells]

14. Wikipedia, 2009.
[Picture: Na+ solid]

15. . Heller, Craig, Orians, Gordon, Purves, William K, Sadava, David, **Life, the Science of Biology**, 6th Edition, (Sinauer Associates, Inc., 2001), 24..
[Diagram: Ionic Bond]

16. Lide, David R., **CRC Handbook of Chemistry and Physics**, 86th Edition, (CRC Press, Taylor & Francis Group, 2005)
[Table: Na+ Properties]

17.   http://www.elementsdatabase.com/.
[Diagram: Periodic Table]

18. Wikipedia, 2009.
[Picture: K+ solid]

19. Lide, David R., **CRC Handbook of Chemistry and Physics**, 86th Edition, (CRC Press, Taylor & Francis Group, 2005).
[Table: K+ properties]

20. Alberts, Bruce, Johnson, Alexander, Lewis, Julian, Raff, Martin, Roberts, Keith, Walter, Peter **Molecular Biology of the Cell**, 4th Edition, (Garland Publishing, Inc., New York, New York, 2002), 1505.
[Picture: RBC membrane thickness]

21. Alberts, Bruce, Johnson, Alexander, Lewis, Julian, Raff, Martin, Roberts, Keith, Walter, Peter **Molecular Biology of the Cell**, 4th Edition, (Garland Publishing, Inc., New York, New York, 2002), 1505.
[Picture: Light and Dark bands of RBC membrane]

22. Alberts, Bruce, Johnson, Alexander, Lewis, Julian, Raff, Martin, Roberts, Keith, Walter, Peter **Molecular Biology of the Cell**, 4th Edition, (Garland Publishing, Inc., New York, New York, 2002), 1505.
[Drawing: Side view of bi-layer of membrane]

23. Alberts, Bruce, Johnson, Alexander, Lewis, Julian, Raff, Martin, Roberts, Keith, Walter, Peter **Molecular Biology of the Cell**, 4th Edition, (Garland Publishing, Inc., New York, New York, 2002), 1505.
[Drawing: 3D structure of the bi-layer]

24. Alberts, Bruce, Johnson, Alexander, Lewis, Julian, Raff, Martin, Roberts, Keith, Walter, Peter , **Molecular Biology of the Cell**, 4th Edition, (Garland Publishing, Inc., 2002.),1516.
[Drawing: Phospholipid Schematic]

25. Alberts, Bruce, Johnson, Alexander, Lewis, Julian, Raff, Martin, Roberts, Keith, Walter, Peter , **Molecular Biology of the Cell**, 4th Edition, (Garland Publishing, Inc., 2002.),1516.
[Drawing: Phosphatidylcholine Formula]

26. Alberts, Bruce, Johnson, Alexander, Lewis, Julian, Raff, Martin, Roberts, Keith, Walter, Peter , **Molecular Biology of the Cell**, 4th Edition, (Garland Publishing, Inc., 2002.),1516.
[Drawing: Phosphatidylcholine van der Waals Space Filling Model]

27. Alberts, Bruce, Johnson, Alexander, Lewis, Julian, Raff, Martin, Roberts, Keith, Walter, Peter , **Molecular Biology of the Cell**, 4th Edition, (Garland Publishing, Inc., 2002), 1524.
[Drawing: 2D Fluid Movement]

28. Alberts, Bruce, Johnson, Alexander, Lewis, Julian, Raff, Martin, Roberts, Keith, Walter, Peter , **Molecular Biology of the Cell**, 4th Edition, (Garland Publishing, Inc., 2002), 1524.
[Drawing: Cholesterol in the bi-layer]

29. Alberts, Bruce, Johnson, Alexander, Lewis, Julian, Raff, Martin, Roberts, Keith, Walter, Peter , **Molecular Biology of the Cell**, 4th Edition, (Garland Publishing, Inc., 2002), 1526.
[Drawing: Schematic, Formula and van der Waals Cholesterol]

30. Alberts, Bruce, Johnson, Alexander, Lewis, Julian, Raff, Martin, Roberts, Keith, Walter, Peter , **Molecular Biology of the Cell**, 4th Edition, (Garland Publishing, Inc., 2002), 1553
[Table: Lipid type and different cell membranes]

31. Alberts, Bruce, Johnson, Alexander, Lewis, Julian, Raff, Martin, Roberts, Keith, Walter, Peter , **Molecular Biology of the Cell**, 4th Edition, (Garland Publishing, Inc., 2002), 1549.
[Diagram: 8 ways a protein associates with the membrane]

32. Alberts, Bruce, Johnson, Alexander, Lewis, Julian, Raff, Martin, Roberts, Keith, Walter, Peter , **Molecular Biology of the Cell**, 4th Edition, (Garland Publishing, Inc., 2002), 1549.
[Diagram: Beta barrel and alpha-helix monolayer]

33. Alberts, Bruce, Johnson, Alexander, Lewis, Julian, Raff, Martin, Roberts, Keith, Walter, Peter , **Molecular Biology of the Cell**, 4th Edition, (Garland Publishing, Inc., 2002), 1553.
[Diagram: alpha-carbon backbone of chain and non-polar side chains]

34. Alberts, Bruce, Johnson, Alexander, Lewis, Julian, Raff, Martin, Roberts, Keith, Walter, Peter , **Molecular Biology of the Cell**, 4th Edition, (Garland Publishing, Inc., 2002), 176.
[Diagram: 4 amino acid polypeptide]

35. . Nelson, David L., Cox, Michael M., **Lehninger Principles of Biochemistry**, 4th Edition, (W.H. Freeman Co., New York, New York, 2004), 121.
[Diagram: Alpha-helix]

35. Nelson, David L., Cox, Michael M., **Lehninger Principles of Biochemistry**, 4th Edition, (W.H. Freeman Co., New York, New York, 2004), 121.
[Diagram: Down the hole of the alpha helix]

36. Alberts, Bruce, Johnson, Alexander, Lewis, Julian, Raff, Martin, Roberts, Keith, Walter, Peter , **Molecular Biology of the Cell**, 4th Edition, (Garland Publishing, Inc., 2002), 1556.
[Diagram: The beta sheet and beta barrel]

37. Guyton, Arthur C., Hall, John E., **Textbook of Medical Physiology**, 11th Edition, (Philadelphia, Elsevier Saunders Inc., 2006), 46.
[Diagram: Chemical content of the ECF and ICF]

38. Guyton, Arthur C., Hall, John E., **Textbook of Medical Physiology**, 11th Edition, (Philadelphia, Elsevier Saunders Inc., 2006), 46.
[Diagram: Physical mechanisms of solute transport across the cell membrane]

39. Guyton, Arthur C., Hall, John E., **Textbook of Medical Physiology**, 11th Edition, (Philadelphia, Elsevier Saunders Inc., 2006), 47.
[Diagram: Gated mechanism of solute transport]

40. Guyton, Arthur C., Hall, John E., **Textbook of Medical Physiology**, 11th Edition, (Philadelphia, Elsevier Saunders Inc., 2006), 47.
[Definition: Na+ channel characteristics]

41. Guyton, Arthur C., Hall, John E., **Textbook of Medical Physiology**, 11th Edition, (Philadelphia, Elsevier Saunders Inc., 2006), 47.
[Definition: K+ channel characteristics]

42. Guyton, Arthur C., Hall, John E., **Textbook of Medical Physiology**, 11th Edition, (Philadelphia, Elsevier Saunders Inc., 2006), 51.
[Diagram: Mechanism of osmosis]

43. Guyton, Arthur C., Hall, John E., **Textbook of Medical Physiology**, 11th Edition, (Philadelphia, Elsevier Saunders Inc., 2006), 52.
[Definition: Normal osmolality]

44. Guyton, Arthur C., Hall, John E., **Textbook of Medical Physiology**, 11th Edition, (Philadelphia, Elsevier Saunders Inc., 2006), 52.
[Definition: Normal osmotic pressure]

45. Guyton, Arthur C., Hall, John E., **Textbook of Medical Physiology**, 11th Edition, (Philadelphia, Elsevier Saunders Inc., 2006), 52.
[Definition: Difference between osmolarity and osmolality]

46. . Guyton, Arthur C., Hall, John E., **Textbook of Medical Physiology**, 11th Edition, (Philadelphia, Elsevier Saunders Inc., 2006), 52.
[Definition: Normal body temperature]

47. Guyton, Arthur C., Hall, John E., **Textbook of Medical Physiology**, 11th Edition, (Philadelphia, Elsevier Saunders Inc., 2006), 53.
[Diagram: Na+ - K+ pump]

48. Kleinsmith, Lewis J., Kish, Valerie M., **Principles of Cell and Molecular Biology**, 2nd Edition, (New York, New York,Harper Collins College Publishers, 1995), 181-2.
[Definition: Terms of energetics formular for ion transport]

49. Guyton, Arthur C., Hall, John E., **Textbook of Medical Physiology**, 11th Edition, (Philadelphia, Elsevier Saunders Inc., 2006), 56.
[Diagram: Tissue transport]

50. Guyton, Arthur C., Hall, John E., **Textbook of Medical Physiology**, 11th Edition, (Philadelphia, Elsevier Saunders Inc., 2006), 292.
[Diagram: Daily Intake and Output of water]

51. Guyton, Arthur C., Hall, John E., **Textbook of Medical Physiology**, 11th Edition, (Philadelphia, Elsevier Saunders Inc., 2006), 292.
[Diagram: Fluid by volume each compartment]

52. Guyton, Arthur C., Hall, John E., **Textbook of Medical Physiology**, 11th Edition, (Philadelphia, Elsevier Saunders Inc., 2006), 294.
[Diagram: ECF and ICF mEq/L]

53. Guyton, Arthur C., Hall, John E., **Textbook of Medical Physiology**, 11th Edition, (Philadelphia, Elsevier Saunders Inc., 2006), 294.
[Diagram: ECF and ICF by osmolar content]

54. Guyton, Arthur C., Hall, John E., **Textbook of Medical Physiology**, 11th Edition, (Philadelphia, Elsevier Saunders Inc., 2006), 182.
[Diagram: Capillary wall]

55. Guyton, Arthur C., Hall, John E., **Textbook of Medical Physiology**, 11th Edition, (Philadelphia, Elsevier Saunders Inc., 2006), 184.
[Table: Permeability of endothelial pore]

56. Guyton, Arthur C., Hall, John E., **Textbook of Medical Physiology**, 11th Edition, (Philadelphia, Elsevier Saunders Inc., 2006), 297.
[Equation: Van't Hoff's Law]

57. . Guyton, Arthur C., Hall, John E., **Textbook of Medical Physiology**, 11th Edition, (Philadelphia, Elsevier Saunders Inc., 2006), 294.
[Table: Osmolarity]

58. . Http://www.thedailyplate.com/nutrition-calories/food, 2009.
[Nutritional Facts]

59. Guyton, Arthur C., **Human Physiology and Mechanisms of Disease,** 4th Edition, (Philadelphia, W.B. Saunders. Co., 1987), 271.
[Transit time]

60. Guyton, Arthur C., **Human Physiology and Mechanisms of Disease,** 4th Edition, (Philadelphia, W.B. Saunders. Co., 1987), 274.
[Magnitude of urine output]

61. http://en.wikipedia.org, 2009.
[Sugar in food forms]

62. http://www.freepatentsonline.com, 2009.
[Osmometer]

63. Guyton, Arthur C., Hall, John E., **Textbook of Medical Physiology**, 11th Edition, (Philadelphia, Elsevier Saunders Inc., 2006), 348.
[Kidney: Range of Dilution]

64. Guyton, Arthur C., Hall, John E., **Textbook of Medical Physiology**, 11th Edition, (Philadelphia, Elsevier Saunders Inc., 2006), 349.
[Renal Response to Diluted Water]

65. Guyton, Arthur C., Hall, John E., **Textbook of Medical Physiology**, 11th Edition, (Philadelphia, Elsevier Saunders Inc., 2006), 349.
[Diagram Nephron]

66. http://upload..wikimedia.org/wikipedia/commons/c/c0/Illu_kidney2.jpg.
[Diagram: Kidney Gross Anatomy]

123

67. http://en.wikipdia.org/wiki/File:Gray 1128.png.
[Diagram: Vasa Recta]

68. http://upload..wikimedia.org/wikipedia/en/2/27/Kidney-medulla.JPG.
[Diagram: Micro Medulla]

69. http://upload..wikimedia.org/wikipedia/en/4/4b/Kidney-Cortex.JPG.
[Diagram: Micro Cortex]

70. Guton, Arthur C., Hall, John E., **Textbook of Medical Physiology**, 11th Edition,
(Philadelphia, Elsevier Saunders Inc., 2006), 350.
[Definition: Maximum concentrating ability of kidney]

71. Guton, Arthur C., Hall, John E., **Textbook of Medical Physiology**, 11th Edition,
(Philadelphia, Elsevier Saunders Inc., 2006), 351.
[Outline: Countercurrent Mechanism]

72. Guton, Arthur C., Hall, John E., **Textbook of Medical Physiology**, 11th Edition,
(Philadelphia, Elsevier Saunders Inc., 2006), 352.
[Table: Countercurrent Mechanism by the #s]

73. Guton, Arthur C., Hall, John E., **Textbook of Medical Physiology**, 11th Edition,
(Philadelphia, Elsevier Saunders Inc., 2006), 353.
[Definition: Urea part in Countercurrent Mechanism]

74. Nelson, David L., Cox, Michael M., **Lehninger Principles of Biochemistry**, 4th
Edition, (W.H. Freeman Co., 2004), 891.
[Diagram: Brain Gross Anatomy]

75. Nelson, David L., Cox, Michael M., **Lehninger Principles of Biochemistry**, 4th
Edition,
(W.H. Freeman Co., 2004), 891.
[Diagram: Axis]

76. Nelson, David L., Cox, Michael M., **Lehninger Principles of Biochemistry**, 4th
Edition,
(W.H. Freeman Co., 2004), 891.
[Diagram: Thalamic nuclei]

77. Nelson, David L., Cox, Michael M., **Lehninger Principles of Biochemistry**, 4th Edition,
(W.H. Freeman Co., 2004), 891.
[Diagram: ADH Molecule]

78. Guyton, Arthur C., Hall, John E., **Textbook of Medical Physiology**, 11th Edition,
(Philadelphia, Elsevier Saunders Inc., 2006), 928.
[Diagram: Thalamic nuclei % Primary Hormone]

79. http://www.freebookcentre.net/medical/neurology.html, 884.
[Diagram: Gross Surface Anatomy Thalamus]

80. http://www.freebookcentre.net/medical/neurology.html, 884.
[Diagram: Cross Section Thirst Center]

81. Weitzman, R.E., Kleeman, C.R., The Clinical Physiology of Water Metabolism-Part 1: The Physiologic Regulation of Arginine Vasopressin Secretion and Thirst, (Medical Progress, Western Journal of Medicine, Nov. 1979), 375.
[Table: Stimuli and Inhibitors of Thirst and ADH]

82.   Dunn, Robert B., USMLE Step 1- Lecture Notes, Physiology, 402.
[Diagram: Negative Feedback Mechanism ADH]

83. Weitzman, R.E., Kleeman, C.R., The Clinical Physiology of Water Metabolism-Part 1: The Physiologic Regulation of Arginine Vasopressin Secretion and Thirst, (Medical Progress, Western Journal of Medicine, Nov. 1979), 496.
[Diagram: ADH Physiologic Function]

84. . Weitzman, R.E., Kleeman, C.R., The Clinical Physiology of Water Metabolism-Part 1: The Physiologic Regulation of Arginine Vasopressin Secretion and Thirst, (Medical Progress, Western Journal of Medicine, Nov. 1979), 496.
[Diagram: Renal Epithelial Surface Changes due to ADH]

85. Weitzman, R.E., Kleeman, C.R., The Clinical Physiology of Water Metabolism-Part 1: The Physiologic Regulation of Arginine Vasopressin Secretion and Thirst, (Medical Progress, Western Journal of Medicine, Nov. 1979), 382.
[Diagram: Magnocellular neuron]

86. Guyton, Arthur C., Hall, John E., **Textbook of Medical Physiology**, 11th Edition,
(Philadelphia, Elsevier Saunders Inc., 2006), 889.
[Definition: Body Temperature Range]

87. Guyton, Arthur C., Hall, John E., **Textbook of Medical Physiology**, 11th Edition, (Philadelphia, Elsevier Saunders Inc., 2006), 890.
[Definition: Total Cardiac Output]

88. Guyton, Arthur C., Hall, John E., **Textbook of Medical Physiology**, 11th Edition, (Philadelphia, Elsevier Saunders Inc., 2006), 891.
[Diagram: Heat Loss]

89. Guyton, Arthur C., Hall, John E., **Textbook of Medical Physiology**, 11th Edition, (Philadelphia, Elsevier Saunders Inc., 2006), 892.
[Definition: Amount of Heat per water weight]

90. Guyton, Arthur C., Hall, John E., **Textbook of Medical Physiology**, 11th Edition, (Philadelphia, Elsevier Saunders Inc., 2006), 893.
[Diagram: Sweat Gland]

91. Guyton, Arthur C., Hall, John E., **Textbook of Medical Physiology**, 11th Edition, (Philadelphia, Elsevier Saunders Inc., 2006), 893, 1065.
[Table: Acclimatization]

92. Guyton, Arthur C., Hall, John E., **Textbook of Medical Physiology**, 11th Edition, (Philadelphia, Elsevier Saunders Inc., 2006), 945.
[Diagram: Adrenal Cortex]

93. Guyton, Arthur C., Hall, John E., **Textbook of Medical Physiology**, 11th Edition, (Philadelphia, Elsevier Saunders Inc., 2006), 947.
[Table: Steroid Activity]

94. Guyton, Arthur C., Hall, John E., **Textbook of Medical Physiology**, 11th Edition, (Philadelphia, Elsevier Saunders Inc., 2006), 950.
[Definition: Aldosterone action time]

95. Godek, S. Fowkes, Bartolozzi, A.R., Godek, J.J., Sweat Rate and Fluid Turnover in American Football Players compared with Runners in a Hot and Humid Environment, Department of Sports Medicine, West Chester University, West Chester, Pennsylvania, 2004,207-8.
[Table: Estimated Fluid Requirements during Sessions]

96. Godek, S. Fowkes, Bartolozzi, A.R., Godek, J.J., <u>Sweat Rate and Fluid Turnover in American Football Players compared with Runners in a Hot and Humid Environment</u>, Department of Sports Medicine, West Chester University, West Chester, Pennsylvania, 2004,207-8.
[Graph: Plasma Change between FB and CC]

97. Godek, S. Fowkes, Bartolozzi, A.R., Godek, J.J., <u>Sweat Rate and Fluid Turnover in American Football Players compared with Runners in a Hot and Humid Environment</u>,

Department of Sports Medicine, West Chester University, West Chester, Pennsylvania, 2004,207-8.
[Graph: Urine Specific Gravity Change between FB and CC]

98. Godek, S. Fowkes, Bartolozzi, Arthur R., Burkholder, Richard, Sugarman, Eric, Peduzzi, Chris, <u>Sweat Rates and Fluid Turnover in Professional Football Players: A Comparison of National Footbass League Linemen and Backs</u>, The H.E.A.T. Institute of West Chester University, West Chester, Pennsylvania, The Journal of Athletic Training, 2008, 185-6.
[Table: Physical Characteristics between Linemen and Backs]

99. Godek, S. Fowkes, Bartolozzi, Arthur R., Burkholder, Richard, Sugarman, Eric, Peduzzi, Chris, <u>Sweat Rates and Fluid Turnover in Professional Football Players: A Comparison of National Footbass League Linemen and Backs</u>, The H.E.A.T. Institute of West Chester University, West Chester, Pennsylvania, The Journal of Athletic Training, 2008, 185-6.
[Graph: Sweat rates and Fluid consumption rates]

100. Godek, S. Fowkes, Bartolozzi, Arthur R., Burkholder, Richard, Sugarman, Eric, Peduzzi, Chris, <u>Sweat Rates and Fluid Turnover in Professional Football Players: A Comparison of National Footbass League Linemen and Backs</u>, The H.E.A.T. Institute of West Chester University, West Chester, Pennsylvania, The Journal of Athletic Training, 2008, 185-6.
[Graph: Sweat Loss, Fluids consumed, and Urine Output]

101. Burkholder, R., Fowkes, Godek S., Peduzzi, C., Condon, S., Kopec, J., Greene, R., <u>Sweat sodium content and sodium losses in NFL players during practices in week 1 versus week 3 of preseason [abstract]</u>, J. Athl. Train., 200r;42 (suppl 2):59S.
[Definition: Na+ rate loss]

Bibliography

Guyton, Arthur C., Hall, John E., <u>Textbook of Medical Physiology</u>, 11th Edition, Elsevier Saunders Inc, 2006

Heller, Craig, Orians, Gordon, Purves, William K, Sadava, David, <u>Life, the Science of Biology</u>, 6th Edition, Sinauer Associates, Inc., 2001

Kotz, John C., Treichel, Paul Jr., <u>Chemistry & Chemical Reactivity</u>, 3rd Edition, Saunders College Publishing, 1996.

Nelson, David L., Cox, Michael M., <u>Lehninger Principles of Biochemistry</u>, 4th Edition, W.H. Freeman Co., 2004.

Alberts, Bruce, Johnson, Alexander, Lewis, Julian, Raff, Martin, Roberts, Keith, Walter, Peter , <u>Molecular Biology of the Cell</u>, 4th Edition, Garland Publishing, Inc., 2002.

Kleinsmith, Lewis J., Kish, Valerie M., <u>Principles of Cell and Molecular Biology</u>, 2nd Edition, Harper Collins College Publishers, 1995.

Godek, S. Fowkes, Bartolozzi, A.R., Godek, J.J., <u>Sweat Rate and Fluid Turnover in American Football Players compared with Runners in a Hot and Humid Environment</u>, Department of Sports Medicine, West Chester University, West Chester, Pennsylvania, 2004.

Godek, S. Fowkes, Bartolozzi, Arthur R., Burkholder, Richard, Sugarman, Eric, Peduzzi, Chris, <u>Sweat Rates and Fluid Turnover in Professional Football Players: A Comparison of National Footbass League Linemen and Backs</u>, The H.E.A.T. Institute of West Chester University, West Chester, Pennsylvania, The Journal of Athletic Training, 2008.

Weitzman, Richard E., Kleeman, Charles R., <u>The Clinical Physiology of Water Metabolism</u>, <u>Medical Progress</u>, The Western Journal of Medicine, 1979.

www.ingramcontent.com/pod-product-compliance
Lightning Source LLC
Chambersburg PA
CBHW081152180526
45170CB00006B/2038